T0394333

Advances in Nuclear Architecture

Niall M. Adams • Paul S. Freemont
Editors

Advances in Nuclear Architecture

Editors
Niall M. Adams
Department of Mathematics
Imperial College London
United Kingdom
n.adams@imperial.ac.uk

Paul S. Freemont
Division of Molecular Biosciences
Imperial College London
United Kingdom
p.freemont@imperial.ac.uk

ISBN 978-90-481-9898-6 e-ISBN 978-90-481-9899-3
DOI 10.1007/978-90-481-9899-3
Springer Dordrecht Heidelberg London New York

Library of Congress Control Number: 2010937564

Printed on acid-free paper

Springer is part of Springer Science+Business Media (www.springer.com)

Foreword

The study of nuclear architecture has benefited considerably from advances in microscopy technology and biochemical techniques in recent years. These advances have enabled both live cell and high-throughput imaging, with a more diverse range of nuclear constituents and perturbations. Disappointingly however, improvements in the capability to visualize nuclei have not been matched by enhancements in the quantitative analysis tools crucial to probing nuclear architecture.

We are of the opinion that careful and rigorous analysis is necessary at all stages of nuclear architecture studies to ensure reliable and reproducible results. In particular, we are concerned that there may sometimes be a preference for describing interesting phenomena in selected images, rather than quantitative description of measurements extracted from 'replicate' images containing multiple nuclei. This preference may arise from the difficulty of obtaining homogeneous replicate images for cells that are growing asynchronously or are in different physiological states. Furthermore experimental conditions for image capture, e.g. cell fixation and indirect immuno-labelling or even over-expression of fluorescent fusion proteins, can cause experimental perturbations leading to heterogeneity and the subsequent challenges of quantitative analysis. However, to make statistically secure statements, replication is essential. Another aspect that is perhaps underplayed is the three-dimensional (3D) nature of nuclei. Over-reliance on 2D representations may be misleading.

The purpose of this book is to provide a snapshot of the current state of nuclear architecture studies, with equal attention on the details of both quantitative analysis and biological outcome. We are extremely fortunate to have been able to gather a diverse group of expert contributors to whom we extend our deepest gratitude for their efforts in making this book what it is.

It is worth commenting on the order of presentation of the contributed chapters. A traditional view might partition the book into parts, related to biological insights versus quantitative analysis, respectively. However, consistent with our vision for the evolution of the field, we prefer to interleave this material to represent the fundamentally inter-disciplinary nature of the problem.

In the first chapter, Graham Dellaire and colleagues focus on three important sub-nuclear domains and their link to cancer, namely PML bodies, nucleoli, and the perinucleolar compartment. They describe the functional roles of sub-nuclear bodies in tumour suppression and oncogenesis, suggesting sub-nuclear domain structure could provide new biomarkers for cancer diagnosis, detection and treatment.

PML bodies are also the discussed by Philip Umande and David Stephens, who illustrate the complexities and power of spatial point pattern analysis by analysing the spatial configuration of PML bodies. This chapter stresses some of the mathematical background which is important to make quantitative analysis reliable.

The theme of quantitative analysis is continued by Christof Cremer and colleagues, who describe quantitative procedures for experimental and modelling approaches for probing human genomic architecture. This includes a 3D computer model of nuclear architecture used to explore chromosome aberrations and quantitative modelling to explore the role of SC35 splicing speckles on nuclear genome structure.

Karl Rohr and co-authors describe sophisticated shape analysis and registration techniques to explore aspects of 3D architecture. The methodology is deployed to analyse the spatial preference of chromatin regions, revealing a variety of interesting relationships.

Joanna Bridger and Ishita Mehta are concerned with the emerging study of the role played by nuclear actin and myosin isoforms as nuclear motor complexes involved in nuclear dynamics. They provide a detailed review of current work, and suggest avenues for future development.

Richard Russell and co-authors return to the theme of quantitative analysis of nuclear architecture, applying methods of spatial analysis to probe the configuration of PML bodies in a variety of conditions. They also suggest an alternative shape analysis procedure to facilitate nuclear registration.

Dean Jackson takes a system view of gene transcription, particularly exploring the relationship between nuclear architecture, at various levels and how they contribute toward the overall control of gene expression. This includes a description of recent work involving gene expression networks during gene induction, and how dynamic behaviour of chromatin fits with current architecture models.

It is clear from the work described in this book that the study of the cell nucleus is reaching an exciting phase where there is an increasing requirement for sophisticated quantitative analyses. This is generating new interdisciplinary research avenues that will ultimately lead to the in silico construction of 'virtual' nuclei. Such advances will not only enable detailed understandings of diseases like cancer but will also allow more fundamental insights where biochemical interactions and transcriptional signalling networks will be integrated with meso-scale spatial models.

London, June 2010

Niall M. Adams
Paul S. Freemont

Contents

Chapter 1
Nuclear Subdomains and Cancer

Kendra L. Cann, Sui Huang, and Graham Dellaire

Abstract Cancer develops when genetic changes, such as the activation of oncogenes and inactivation of tumour suppressors, allow a cell to escape the normal growth and proliferation restrictions. These functional changes ultimately result in structural alterations at both the nuclear and cellular levels. As such, cell morphology and biological marker expression are some of the main criteria in tumour pathology for diagnosis and prognosis. The development of advanced microscopy techniques has provided a much more detailed map of the nuclear landscape, and because of this, structural changes in subnuclear bodies induced during oncogenesis can be readily visualized. This type of analysis has identified novel cancer biomarkers in the form of nuclear structures associated with malignancy, such as the perinucleolar compartment (PNC). It has also allowed a much more detailed examination of nuclear body function, which has provided novel mechanisms and regulators of tumour suppression and oncogenesis. This chapter will focus on three of the most important subnuclear domains for cancer biology: promyelocytic leukemia nuclear bodies (PML NBs), the nucleolus, and the perinucleolar compartment (PNC).

Keywords Promyelocytic leukemia bodies (PML NBs) • Nucleolus • Perinucleolar compartment (PNC) • Cancer • p53 • Tumour suppressor • Oncogene • Biomarkers

K.L. Cann and G. Dellaire (✉)
Department of Pathology, Dalhousie University, Rm 11G1, Sir Charles Tupper Medical Building, 5859 University Avenue, NS B3H 4 H7, Halifax
and
Beatrice Hunter Cancer Research Institute, Halifax, Nova Scotia, Canada
e-mail: dellaire@dal.ca

S. Huang
Northwestern University, Feinberg School of Medicine, Cell and Molecular Biology, Chicago, Illinois 60611, USA

N.M. Adams and P.S. Freemont (eds.), *Advances in Nuclear Architecture*,
DOI 10.1007/978-90-481-9899-3_1, © Springer Science+Business Media B.V. 2011

Abbreviations

AgNORs	Silver stained nucleolar organizer regions
AKT	V-akt murine thymoma viral oncogene homolog 1
ALT	Alternative lengthening of telomeres
APL	Acute promyelocytic leukaemia
ARF	Alternate reading frame protein
ASK1	Mitogen-activated protein kinase kinase kinase 5
ATM	Ataxia telangiectasia mutated
ATR	ATM- and Rad3-related
ATRIP	ATR-interacting protein
Bcl-2	B-cell CLL/lymphoma 2
BCL2L11	Bim/BCL2-like 11
Bop1	Block of proliferation 1
BLM	Bloom syndrome protein
Brca1	Breast cancer 1
Br-U	Bromouridine
CASP8AP2	Caspase 8 associated protein 2
CBP	CREB binding protein
CDK	Cyclin-dependent kinase
CK1	Casein kinase 1
CKII	Casein kinase 2
CUG-BP	CUG triplet repeat, RNA binding protein 1
DAXX	Death-domain associated protein
DBA	Diamond-Blackfan anemia
DC	Dyskeratosis congenital
DKC1	Dyskeratosis congenital, dyskerin
DFC	Dense fribrillar component
E6AP	E6-associated protein
ERK2	Extracellular-signal-regulated protein kinase 2
FADD	Fas (TNFRSF6)-associated via death domain
FCs	Fibrillar centres
FLASH	Flice-associated huge protein
FOXO	Forkhead box
γ-H2A.X	Phosphorylated H2A.X
GC	Granular component
HAUSP	Herpes-associated ubiquitin-specific protease
HDAC1	Histone deacetylase 1
HIC1	Hypermethylated in cancer 1
HIPK2	Homeodomain interacting protein kinase 2
hMSH6	Human mutS homolog 2
HP1	Heterochromatin protein 1
HPV	Human papilloma virus
HSV1	Herpes Simplex Virus 1
IAPs	Inhibitor of apoptosis proteins

IFNs	Interferons
IGF-1	Insulin-like growth factor 1
IkB	Inhibitor of NF-kB
ING-1	Inhibitor of growth 1
IRES	Internal ribosome entry site
IRS	Insulin receptor substrate
JNK	c-Jun-N-terminal kinase
Kap1	KRAB domain-associated protein 1
KSRP	KH-type splicing regulatory protein
KSHV	Karposi sarcoma-associated herpesvirus
LANA2	Latency associated nuclear antigen 2
MAPK	Mitogen-activated protein kinase
MDM2	Mouse double minute 2
Mina53	MYC induced nuclear antigen
Misu	Myc-induced SUN domain-containing protein
Mre11	Meiotic recombination 11
MRN	Mre11, Rad50, NBS1
mRNA	Messenger RNA
MTA2	Metastasis-associated protein 2
mTOR	Mammalian target of rapamycin
NBS1	Nijmegen breakage syndrome 1
Nop2	Nucleolar protein homolog (yeast) 2
NORs	Nucleolar organizer regions
NPM	Nucleophosmin
NF-κB	Nuclear factor of kappa light polypeptide gene enhancer in B-cells 1
OPT	Oct1, PTF transcription
PAWR	PRKC, apoptosis, WT1, regulator
PI3K	Phosphatidylinositol 3-kinase
PODs	PML oncogenic domains
PML NBs	Promyelocytic leukemia nuclear bodies
PNC	Perinucleolar compartment
PP2A	Protein phosphatase 2
pRB	Retinoblastoma protein
pol	Polymerase
PRMT1	Protein arginine methyltransferase 1
PTEN	Phosphatase and tensin homolog
PTB	Polypyrimidine track binding protein
PTF	PSE-binding factor subunit delta,small nuclear RNA activating complex, polypeptide 2, 45kDa
RanBP2	Ran Binding Protein 2
RARα	Retinoic acid receptor alpha
RBCC	RING zinc finger, two B-boxes and coiled-coil domain
rDNA	Ribosomal DNA
RecQL4	RecQ protein-like 4
RING	Really interesting new gene

RMRP	RNA component of mitochondrial RNA processing endoribonuclease
RNF4	Ring finger protein 4
RPLs	Ribosomal proteins of the large subunit
RPSs	Ribosomal proteins of the small subunit
rRNA	Ribosomal RNA
RSK	Ribosomal protein S6 kinase, RPS6KA1
SAHF	Senescence-induced heterochromatin foci
scaRNA	Small Cajal body RNA
SCE	Sister-chromatid exchange
SENPs	SUMO1/sentrin specific peptidase 1
SFCs	Splicing factor compartments
SIM	SUMO-interacting motif
SIRT1	Sirtuin (silent mating type information regulation 2 homolog) 1
snRNPs	Small nuclear ribonucleoproteins
SRP	Signal recognition particle
SUMO	Small ubiquitin modifier
TAF1	TBP-associated factor 1
TCAB1	Telomerase Cajal body protein 1
TERC	Telomerase RNA component
TERT	Telomerase reverse transcriptase
TGF-β	Transforming growth factor beta
THAP1	THAP domain containing apoptosisassociated protein 1
TIF-IA	Transcription initiation factor IA
TIF-IB	Transcription initiation factor IB
TIP60	HIV-1 Tat interactive protein, KAT5 K(lysine) acetyltransferase 5
TNFα	Tumour necrosis factor alpha
TNFSF10	TRAIL/TNF superfamily, member 10
TopBP1	Topoisomerase (DNA) II binding protein 1
TRF2	Telomere repeat binding factor 2
TSA	Trichostatin A
TTF-1	Transcription terminator factor-1
SL1	Promoter selectivity factor
UBC9	Ubiquitin-conjugating enzyme E2I(UBC9 homolog, yeast)
UBF	Upstream binding factor
USP7	Ubiquitin specific protease 7
WRN	Werner syndrome, RecQ helicase-like
YAP	Yes-associated protein
ZIPK	ZIP kinase, death-associated protein kinase 3

1.1 Introduction

The main criterion for cancer diagnosis and prognosis is the pathological evaluation of tumour tissue and/or cancer cells. This type of examination relies heavily on the morphology of and biological markers expressed by the cancer cells, which provide information on the cell of origin, differentiation status, and proliferation potential. Determining the proliferation potential of a tumour is important for identifying the aggressiveness of a given cancer, and because of this, the nucleus has garnered an important place in tumour pathology (Elias 1997). However, because the origins of any cancer lie in the genetic changes it contains, the importance of the nucleus to cancer biology cannot be underscored too strongly. These genetic changes lead to alterations in gene and protein expression, inducing functional and ultimately structural changes within the cell that together facilitate carcinogenesis to occur.

The relationship between structure and function pervades biology, and the nucleus is no exception. As the storehouse of a cell's genetic information, the nucleus performs many diverse and critical functions, including DNA replication, DNA damage sensing and repair, cell cycle regulation, control of apoptosis, RNA transcription (by RNA polymerases I, II, and III), RNA processing, ribonucleoprotein complex formation, and induction of cell senescence, all of which can and do function in carcinogenesis (Dundr and Misteli 2001; Balmain et al. 2003; Raska et al. 2006; Dellaire and Bazett-Jones 2007). The nuclear landscape is organized around these events, and as our knowledge of nuclear structure improves so too does our understanding of the processes themselves and how they are integrated and regulated.

The nucleus is first organized around the DNA itself. In humans, a nucleus of approximately 10 μm contains about 2 m of DNA, distributed among 46 chromosomes (Dundr and Misteli 2001). The DNA in each chromosome is first compacted at the level of the nucleosome (Van Holde et al. 1980), and then this is further twisted and folded multiple times to generate chromosomes, which reside within the nucleus in distinct chromosome territories (Gilbert et al. 2005). Within the chromosomes, the chromatin is organized into regions of facultative and constitutive heterochromatin (which are highly compacted and transcriptionally inactive) and euchromatin (which is more relaxed and transcriptionally active) (Delcuve et al. 2009). Interacting with the chromatin are numerous proteinaceous nuclear substructures (Fig. 1.1), the largest of which are the nucleoli. The other subnuclear organelles are the promyelocytic leukemia nuclear bodies (PML NBs), nuclear speckles (splicing factor compartments, interchromatin granule clusters), Cajal bodies (coiled bodies), OPT (Oct1, PTF transcription) domains, gems, cleavage bodies, Polycomb Group bodies, Sam68 nuclear bodies, and the perinucleolar compartment (PNC) (Dundr and Misteli 2001; Spector 2001; Dellaire and Bazett-Jones 2007). Polycomb Group bodies are associated with pericentromeric heterochromatin, so they may be involved in gene silencing (Spector 2001). Nuclear speckles are thought to function in the assembly, maturation, and storage of snRNPs (small nuclear ribonucleoproteins), which function in messenger RNA (mRNA) splicing (Handwerger and Gall 2006). OPT domains are associated primarily with transcription factors, including PTF, Oct1, TBP, and Sp1, as well as RNA polymerase II (Pombo et al. 1998). Cajal bodies, of which there are 1–10 per cell, function in snRNP modification and assembly, as well as snoRNP (small nucleolar ribonucleoprotein) modification (Morris 2008). These modifications include methylations and

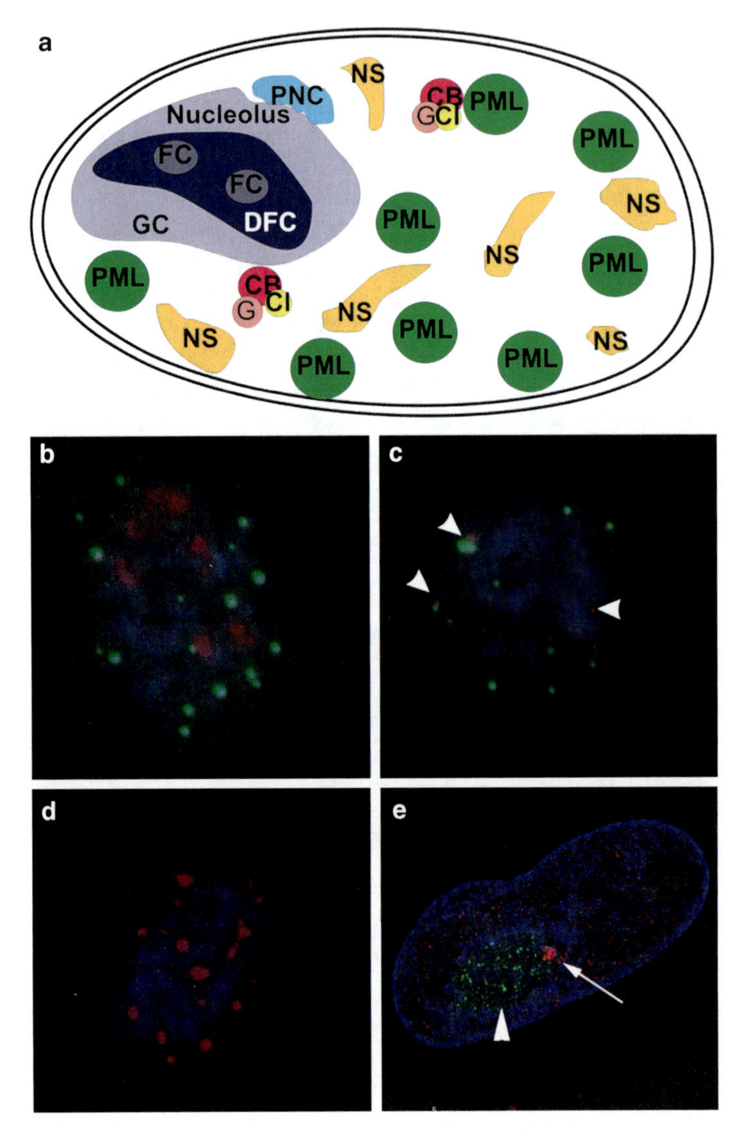

Fig. 1.1 The subnuclear bodies. (**a**) Cartoon depicting a landscape map of some of the nuclear bodies. Structure of the nucleolus: FC, fibrillar centre; DFC, dense fibrillar component; GC, granular component. PNC, perinucleolar compartment; PML, PML NBs; CB, Cajal bodies; G, gem; Cl, cleavage body; NS, nuclear speckle. (**b**) Immunofluorescence image of nucleoli and PML NBs. A normal human diploid fibroblast (NHDF) cell was stained with antibodies to nucleolin (*red*) to mark the nucleoli, and PML (*green*) to mark the PML NBs. (**c**) Immunofluorescence image of Cajal bodies and PML NBs. A NHDF cell was stained with antibodies to PML (*green*) for the PML NBs, and coilin (*red*) to visualize to the Cajal bodies (*arrowheads*). (**d**) Immunofluorescence image of nuclear speckles. A NHDF cell was stained with an antibody to SC35, a marker protein of nuclear speckles. (**e**) Immunofluorescence image of a PNC. The cell was immunolabeled with anti-fibrillarin antibody (*green*), which marks the nucleolus (*arrowhead*) and is co-labeled with anti-PTB antibody (*red*), which stains the PNC (*arrow*). The bar = 5 um. In all images, the cells have been countered stained with DAPI (*blue*) to visualize the DNA

pseudouridylations that are specified by the scaRNAs (small Cajal body RNAs) (Morris 2008). Gems (gemini of Cajal bodies) colocalize with or are adjacent to Cajal bodies. These structures are enriched for SMN (survival of motor neurons), which is critical for the transport of snRNPs from the cytoplasm to the Cajal bodies (Morris 2008). Next, cleavage bodies, which also usually overlap with or lie adjacent to Cajal bodies, contain factors involved in the cleavage and polyadenylation steps of mRNA processing (Spector 2001). Therefore, the above mentioned structures mainly function in RNA polymerase II-dependent transcription and mRNA processing.

The nucleolus is the site of RNA polymerase I-dependent rRNA (ribosomal RNA) transcription and ribosome biogenesis. However, it also functions in the processing and maturation of non-nucleolar RNA and ribonucleoproteins, and in mRNA export. Furthermore, it has an important regulatory role with respect to the cell cycle, DNA replication, DNA repair, cell senescence, tumour suppression, and the cell stress response (Maggi and Weber 2005; Raska et al. 2006; Boisvert et al. 2007; Dellaire and Bazett-Jones 2007; Montanaro et al. 2008; Sirri et al. 2008). Like the nucleolus, PML NBs are plurifunctional, and function in gene transcription, tumour suppression, apoptosis, cell senescence, and DNA repair (Dellaire and Bazett-Jones 2004; Bernardi and Pandolfi 2007). Finally, Sam68 nuclear bodies and the PNC are nuclear structures that are associated with the development of malignancy, as they are rarely observed in primary cells (Huang et al. 1997; Huang 2000; Norton et al. 2008a). They are also both associated with the nucleolus, and are thought to function in RNA metabolism (Huang 2000; Pollock and Huang 2009).

All of these substructures lack membranes, but they all contain defining sets of marker proteins, they can be morphologically identified by light and/or electron microscopy, and some of them can be biochemically purified (Dundr and Misteli 2001). Because these nuclear bodies lack membranes, how they exist as distinct entities within the nucleus is not completely understood; however, it is clear that ongoing cellular processes (such as rRNA transcription for the nucleolus) and self-associatio of proteins (such as for the PML protein in PML NBs and coilin in Cajal bodies) are capable of seeding the assembly of many of these bodies (Hernandez-Verdun 2006a; Rippe 2007; Emmott and Hiscox 2009; Matera et al. 2009). As described above, most of these structures are involved in nuclear processes such as RNA transcription, splicing, and/or ribonucleoprotein assembly, and some also make physical contacts with chromatin, such as with the nucleolus and PML NBs (Dellaire and Bazett-Jones 2004; Raska et al. 2006). Because of this, some of these bodies undergo dramatic reorganizations during S phase when the DNA is being replicated (Dellaire et al. 2006b) and/or during mitosis (Sirri et al. 2008). Furthermore, most also respond to system perturbation, such as during transcriptional inhibition, DNA damage, cell stress, growth factor induction, and apoptosis. It is perhaps not surprising then that tumour suppressors and oncoproteins that affect these processes can also affect the structure and stability of these bodies (Dundr and Misteli 2001; Raska et al. 2006; Dellaire and Bazett-Jones 2007; Derenzini et al. 2009).

Nuclear bodies can also function beyond RNA metabolism, and, as described above, directly affect pathways such as the growth regulation, DNA repair, cell cycle checkpoint regulation, the cell stress response, apoptosis, tumour suppression, telomere maintenance, and senescence (Dellaire and Bazett-Jones 2004; Raska et al. 2006; Bernardi and Pandolfi 2007; Boisvert et al. 2007; Dellaire and Bazett-Jones 2007; Salomoni et al. 2008; Sirri et al. 2008). Therefore, nuclear bodies are significant for cancer biology because (1) they can be physically affected by changes occurring during carcinogenesis, allowing them to be used as cancer biomarkers, (2) their functions are directly related to tumour suppression and mutations in their constituent proteins can impact oncogenesis, and (3) their normal functions can be harnessed to fuel malignancy (Maggi and Weber 2005; Bernardi and Pandolfi 2007; Dellaire and Bazett-Jones 2007; Montanaro et al. 2008; Derenzini et al. 2009). This review will focus on three of the most important nuclear bodies for cancer biology: PML NBs, nucleoli, and the perinucleolar compartment. PML NBs and nucleoli are normal nuclear structures that provide critical tumour suppressive, and, in the case of the nucleolus, oncogenesis-supporting functions. On the other hand, the PNC represents a nuclear structure that is acquired during malignant adaptation. However, it is important to note that none of these nuclear bodies exist in isolation. As will be discussed at the end of the chapter, just as their protein and RNA constituents are in constant flux, trafficking between nuclear substructures in coordination with changes in cell state, so too are their functions coordinated in tumour suppression and oncogenesis.

1.2 Promyelocytic Leukemia Nuclear Bodies and Tumour Suppression

The promyelocytic leukemia (PML) nuclear body (NB) is a protein-rich subnuclear domain first identified by anti-nuclear antibodies in the serum of patients with primary biliary cirrhosis (Szostecki et al. 1990; Sternsdorf et al. 1995). These bodies are also referred to in the literature as PML oncogenic domains (PODs), nuclear domain 10 (ND10) or PML nuclear domains (PML-NDs). PML NBs have been studied predominately in mammalian cells, they number between 10–30 bodies per cell and range from 0.3–1.0 μm in diameter (Fig. 1.1). PML NBs are highly dynamic, changing in number and biochemical composition during the cell cycle and in response to cellular stress. Although more than 70 proteins have been found to localise to the PML NB (see: Nuclear Protein Database (Dellaire et al. 2003), the main structural component is the PML protein. Although no identifiable homologues of the PML protein have been found in non-vertebrate species or single cell eukaryotes, the PML protein is expressed in several vertebrate species including mammals, marsupials and the chicken.

The gene encoding PML was initially identified at the break point of a common translocation, t(15;17), associated with the development of acute promyelocytic leukemia (APL), that juxtaposes the *PML* gene on chromosome 15 with the retinoic acid receptor alpha (*RARα*) on chromosome 17 (de The et al. 1991; Kakizuka et al. 1991). As a consequence a potent oncogene is created by the fusion of these gene sequences, and the expression of the PML-RARα protein disrupts PML NB formation as well as altering the cell survival and differentiation of promyelocytes, which leads to the development of APL. Treatment of APL with retinoic acid or arsenic trioxide triggers the degradation of the PML-RARα protein, which in turn leads to the reformation of PML NBs and remission of the leukemia by restoring normal differentiation of promyelocytes during hematopoiesis (Zhu et al. 2001). A role for PML in tumour suppression of other malignancies is supported by the observation of reduced PML protein levels in solid tumours of diverse histological origin (Gurrieri et al. 2004) and the increased susceptibility of PML knock-out mice to cancer following treatment with carcinogens (Wang et al. 1998b). Reduced PML NB number correlates with the development of prostate and colon cancer (Gurrieri et al. 2004) and PML knock-out mice are prone to prostate hyperplasia, a hallmark of the initiation of prostate cancer (Trotman et al. 2006). In addition, recent evidence also suggests that loss of the PML gene expression is also associated with hyperproliferation of stem cells in the brain and the mammary gland, and that PML expression is required for the maintenance of stem cell-like Leukemia Initiating Cells (LICs) (reviewed in [Salomoni 2009]).

PML NB structure and composition is believed to play a role in determining the cellular function of these bodies. Their protein composition is regulated by expression of different isoforms of the PML protein as well as the post-translational modification of these isoforms by the small ubiquitin modifier (SUMO). At least seven canonical isoforms of PML, expressed from a single gene (PML I through VII), have been characterised (Jensen et al. 2001). These isoforms contain a highly conserved invariant N-terminus consisting of a Really Interesting New Gene (RING) zinc finger, two B-boxes and coiled-coil domain (RBCC), encoded by the first four exons, followed by alternative C-terminal exons. These isoforms exhibit distinct protein interactions, and when over-expressed, produce morphologically distinct PML NBs (Condemine et al. 2006). All of the isoforms with the exception of the cytoplasmic isoform PML-VII, contain a nuclear localisation signal (NLS) encoded within exon 6. PML NB formation and the localisation of several canonical nuclear body components, including SP100 and DAXX, requires the post-translational modification of PML by SUMO (Ishov et al. 1999; Zhong et al. 2000a; Shen et al. 2006). PML is sumoylated on several lysine residues (K65 in the Ring finger domain, K160, and K490 within the NLS) by SUMO isoforms 1–3 (Sternsdorf et al. 1997; Fu et al. 2005). The sixth exon of PML isoforms I-VI also contains a SUMO-interaction motif (SIM). The hetero-oligomerization of different isoforms of PML, via binding of the SIM of one isoform to the SUMO-modified lysines of other PML isoforms, contributes to PML NB formation (Shen et al. 2006).

In addition to tumour suppression, PML NBs have been implicated in a number of cellular functions including host viral defense, the post-translational modification and proteolysis of proteins, transcriptional and post-transcriptional gene regulation, DNA damage signaling, telomere maintenance, cell senescence and apoptosis (reviewed in [Dellaire and Bazett-Jones 2004; Bernardi and Pandolfi 2007]). It is believed that the PML NB may contribute to these processes, most of which play a role in tumour suppression, by sequestering or releasing protein factors to and from the nucleoplasm, and/or facilitating their post-translational modification or degradation. As such, the PML NB may represent an important cellular node within the nucleus for the integration of complex signaling pathways that together maintain appropriate cell growth and differentiation. In this chapter, we will summarize the evidence supporting a multi-faceted role for PML and PML NBs in tumour suppression.

1.2.1 PML Plays a Role in Multiple Tumour Suppressor Pathways

There are many parallels that can be drawn between the classical tumour suppressor p53 and PML. Similar to *TP53* knock-out mice, at first glance *PML* null mice appear healthy and viable until challenged with carcinogens or radiation, at which time a propensity for the development of neoplasms becomes apparent (Harvey et al. 1993; Kemp et al. 1994; Wang et al. 1998a). Like p53, PML appears to be quite promiscuous in its protein-protein interactions and both proteins have been implicated in transcriptional regulation, DNA damage signaling, apoptosis, and in cell senescence (Dellaire and Bazett-Jones 2004; Bernardi and Pandolfi 2007; Junttila and Evan 2009; Levine and Oren 2009). Finally, both proteins are modified by SUMOylation (Sternsdorf et al. 1997; Gostissa et al. 1999), the levels of p53 and PML are tightly regulated through ubiquitin-mediated protein degradation (Scheffner et al. 1993; Lallemand-Breitenbach et al. 2008; Tatham et al. 2008), and reduced levels of these factors are associated with the development of cancer in multiple tissues (Chang et al. 1993; Gurrieri et al. 2004). PML and PML NBs are also directly implicated in the regulation of p53 and a number of other tumour suppressor pathways, summarised in Table 1.1 (reviewed in [Salomoni and Pandolfi 2002; Salomoni et al. 2008]). In this section, we will focus on the regulation of p53 by PML, which serves as a useful paradigm for understanding the possible mechanisms by which this nuclear body might regulate different tumour suppressor pathways.

1.2.1.1 The Regulation of p53 Function by PML: A Paradigm for PML NB Function in Tumour Suppression

The tumour suppressor p53 is an important regulator of the cell cycle that is often found mutated in all forms of cancer. Loss of p53 function impairs the ability of the cell to make cell fate decisions following genotoxic stress including initiating cell cycle arrest and apoptosis. Normally, levels of p53 are kept low by

Table 1.1 Tumour suppressor pathways associated with PML and PML NBs

Pathway	Protein	Notes	Reference
Cell cycle regulation and cellular senescence			
	p53	Act, Exp, IP, NB, Stbl	Pearson et al. 2000; Ferbeyre et al. 2000; Guo et al. 2000; D'Orazi et al. 2002; Hofmann et al. 2002
	pRb	IP, NB	Alcalay et al. 1998
	SIRT1	NB, IP	Langley et al. 2002
DNA damage signaling and apoptosis			
	ATM	NB	Stagno D'Alcontres et al. 2007
	ATR	NB[a], P	Barr et al. 2003; Bernardi et al. 2004
	BLM	NB	Ishov et al. 1999
	Chk2	Act, IP, NB, P	Yang et al. 2002; Yang et al. 2006
	CK1	IP	Alsheich–Bartok et al. 2008
	CK2	P, Deg	Scaglioni et al. 2006
	DAXX	IP, NB	Ishov et al. 1999
	FLASH	NB	Milovic–Holm et al. 2007
	HAUSP	Deg[b], NB	Everett et al. 1997; Everett et al. 1998
	HIPK2	NB, P	D'Orazi et al. 2002; Hofmann et al. 2002; Gresko et al. 2009
	Mdm2	IP, NB	Kurki et al. 2003; Bernardi et al. 2004
	MRE11	Act, Meth, NB[c]	Lombard and Guarente 2000; Boisvert et al. 2005
	NBS1	NB[a, c]	Lombard and Guarente 2000; Wu et al. 2000; Naka et al. 2002
	p63	Act, Stbl	Bernassola et al. 2005
	p73	Act, Stbl	Bernassola et al. 2004
	Par-4	NB	Kawai et al. 2003; Roussigne et al. 2003
	PIAS4	NB[d]	Sachdev et al. 2001; Sun et al. 2005
	Pin1	Deg, IP, NB[d]	Reineke et al. 2008
	Rad50	NB[a, c]	Xu et al. 2003
	RASSF1C	NB[d]	Kitagawa et al. 2006
	RPA	NB[a, d]	Barr et al. 2003; Dellaire et al. 2006b
	THAP1	NB	Roussigne et al. 2003
	TIP60	NB, Stbl	Cheng et al. 2008; Wu et al. 2009
	TopBP1	NB, Stbl	Xu et al. 2003
	YAP	NB, Sumo	Strano et al. 2005; Lapi et al. 2008
	ZIPK	NB	Kawai et al. 2003
Regulation of protein synthesis			
	mTOR	NB	Bernardi et al. 2006
	eIF4E	IP, NB[d, f]	Lai and Borden 2000; Scheper et al. 2003
TGF-β signaling			
	SARA	IP, CB	Lin et al. 2004
	SMAD2/3	IP, CB	Lin et al. 2004
	SnoN	IP, NB	Pan et al. 2009
PI3/PTEN/AKT signaling			
	PTEN	NB	Song et al. 2008
	AKT	IP, NB[e]	Trotman et al. 2006

(continued)

Table 1.1 (continued)

Pathway	Protein	Notes	Reference
NF-κB			
	RelA/p65	IP, NB[d]	Wu et al. 2003
Wnt/β-catenin signaling			
	β-catenin	IP, NB[d]	Shtutman et al. 2002

Notes: *Act* = activation regulated by PML; *CB* = colocalises with cytoplasmic PML bodies; *Deg* = controls degradation of PML; *Exp* = controls PML gene expression; *IP* = immunoprecipitated with PML; *NB* = localised to PML NBs; *Meth* = arginine methylated at PML NBs; *P* = phosphorylates PML; *Stbl* = protein levels stabilized by PML; *Sumo* = Sumoylation promoted by PML; [a]in ALT cells or after DNA damage; [b]in the context of herpes virus infection; [c]cell cycle-dependent; [d]partial colocalisation or juxtapositioning with PML NBs; [e]phospho-form of AKT; [f]association with PML NBs may be antibody or fixation-dependent

ubiquitin-mediated proteolytic degradation through the action of Mdm2, which is an E3 ligase that ubiquitinates p53 (reviewed in [Kruse and Gu 2009]). When the DNA of a cell is damaged, p53 is no longer degraded and accumulates in the cell, in addition p53 is post-translationally modified promoting its ability to act as a transcriptional activator of important genes involved in cell cycle regulation (e.g. the CDK2 inhibitor p21) and apoptosis (e.g. Puma, Bax and Noxa) (Menendez et al. 2009). PML knock-out mice show impaired activation of p53 in response to genotoxic stress (Guo et al. 2000). As we will discuss below, this phenotype can partly be explained by the fact that PML and PML NBs play an important role in both the post-translational modification and stabilization of p53.

The earliest studies to identify a regulatory role for PML in p53 function involved the study of Ras-induced senescence (Ferbeyre et al. 2000; Pearson et al. 2000). When cells express an oncogenic form of the Ras GTPase (e.g. Ras V12), both PML and p53 protein levels increase resulting in the induction of p53 response genes such as p21 that arrest cell growth and contribute to senescence (Ferbeyre et al. 2000). Acetylation and activation of p53 in this context requires PML NBs, which act as sites for the co-accumulation of p53 and its acetyltransferase CBP/p300. Ras-induced senescence can be blocked by the expression of the adenovirus oncoprotein E1A, which disrupts PML NBs and prevents p53 phosphorylation, providing further evidence that PML NBs are required for p53 activation in response to oncogenic Ras (Ferbeyre et al. 2000).

PML NBs are also sites of the accumulation and post-translational modification of p53 following UV-irradiation (D'Orazi et al. 2002; Hofmann et al. 2002). In this context, p53 is recruited to PML NBs with CBP and HIPK2, which act together to activate p53 by acetylation on Lys382 and phosphorylation on Ser46, respectively. Acetylation of p53 at Lys382 is antagonized by the SIRT1, a homologue of the yeast NAD-dependent deacetylase known as silent information regulator 2 (Sir2) (Langley et al. 2002). SIRT1 localises to PML NBs and inhibits oncogenic Ras-mediated senescence by inhibiting the activation of p53 by deacetylating Lys382. Lys382 of p53 may also be targeted by HDAC1, which is recruited to deacetylate p53 via both the KRAB domain-associated protein 1 (Kap1) (Wang et al. 2005) and the metastasis-associated protein 2 (MTA2) (Luo et al. 2000). Another protein acetylase, Tat-interactive protein 60 (TIP60), also localises to PML NBs following UV-mediated DNA damage (Cheng et al. 2008). UV-irradiation promotes TIP60 SUMOylation, which in turn increases the acetyltransferase activity of TIP60 and leads to its association with PML NBs (Cheng et al. 2008). TIP60 is able to acetylate Lys120 of p53, a modification that was shown to facilitate p53-dependent apoptosis (Sykes et al. 2006; Yang et al. 2006). The association of TIP60 with PML also protects the acetyltransferase from ubiquitination and proteasomal degradation by disrupting its interaction with Mdm2 (Wu et al. 2009). Thus, p53 function is regulated by acetylation at PML NBs by both CBP and TIP60.

The recruitment of p53 to PML NBs is mediated in part by the interaction of PML with the DNA binding domain of p53 (Fogal et al. 2000; Guo et al. 2000). Although when over-expressed in normal human fibroblasts several isoforms of PML can recruit p53 to PML NBs (i.e. PML I-V) (Bischof et al. 2002), PML isoform IV binds most efficiently to p53 (Guo et al. 2000) and is primarily responsible for stabilizing and activating the tumour suppressor (Bischof et al. 2002). However, the over-expression of PML IV alone in PML –/– murine embryonic fibroblasts fails to induce senescence (Bischof et al. 2002). These data serve to highlight the importance of PML isoforms in regulating cell growth and imply that hetero-oligomerization of PML IV with other isoforms of PML is a requirement for the post-translational modification and activation of p53.

PML can also regulate the p53 protein by a second mechanism, which involves inhibiting the ubiquitination of p53 by the E3-ligase Mdm2 (Kurki et al. 2003). PML can interact with both p53 and Mdm2, and therefore it is thought that either the formation of a trimeric complex between these proteins inhibits Mdm2-mediated ubiquitination of p53, or that PML sequestration of Mdm2 prevents it interaction with p53, for example at the nucleolus following DNA damage (Kurki et al. 2003; Bernardi et al. 2004). Alternately or in addition, PML may inhibit the ability of Mdm2 to target p53 by promoting the phosphorylation of p53 by Chk2 and CK1 on Ser20 and Thr18, respectively (Louria-Hayon et al. 2003; Alsheich-Bartok et al. 2008). Although it is unclear if PML directly affects the enzymatic activity of CK1, Chk2 autophosphorylation and activation is mediated by PML (Yang et al. 2006), and the co-recruitment of p53 and Chk2 at PML NBs could facilitate p53 phosphorylation (Louria-Hayon et al. 2003).

Recently, yet another mechanism of p53 stabilization by PML has emerged involving the ubiquitin-specific protease 7 (USP7), also known as the herpes-associated ubiquitin-specific protease (HAUSP) (Everett et al. 1997). PML was shown to promote the de-ubiquitination of p53 by USP7 (Li et al. 2002). USP7 is found to colocalise to a subset of PML NBs (Everett et al. 1997) and thus the co-recruitment of USP7 and p53 at PML NBs could mediate the de-ubiquitination and stabilization of p53. In addition, under steady-stated conditions USP7 can form a ternary complex with Mdm2 and the PML NB-associated protein DAXX, enhancing Mdm2 stability (Li et al. 2004; Tang et al. 2006).

Following DNA damage, Mdm2 dissociates from DAXX and levels of Mdm2 decrease as its ubiquitinated form accumulates and is degraded by the proteasome (Tang et al. 2006).

The examples of p53 regulation by PML described above highlight several general mechanisms by which PML and PML NBs contribute to the regulation of tumour suppression. The first is the sequestration and/or post-translational modification of tumour suppressors by enzymes that co-accumulate at PML NBs. For example, PML modulates the acetylation and stabilization of both the p53 and p73 tumour suppressors (D'Orazi et al. 2002; Hofmann et al. 2002; Bernassola et al. 2004), the dephosphorylation of activated AKT (Trotman et al. 2006), the arginine methylation and activation of the DNA repair protein MRE11 (Boisvert et al. 2005), and the sumoylation and stabilization of the breast tumour suppressor known as the Yes-associated protein (YAP) (Lapi et al. 2008; Yuan et al. 2008). A second mechanism involves the modulation of tumour suppressor protein levels through modulation of ubiquitin-mediated proteasomal degradation. For example, PML can disrupt protein interactions between the ubiquitin E3 ligase Mdm2 and p53 (Kurki et al. 2003; Bernardi et al. 2004), and inhibits the proteasomal degradation of p63, p73 and YAP (Bernassola et al. 2004; Bernassola et al. 2005; Lapi et al. 2008). A third mechanism involves PML isoform-specific recruitment to PML NBs and/or activation of a given tumour suppressor, which enables fine-tuning of tumour suppressor regulation irrespective of PML protein levels through alternative splicing of the PML gene; as exemplified by the specific activation of p53 by PML isoform IV (Fogal et al. 2000; Guo et al. 2000; Bischof et al. 2002). Finally, in some cases there exists an additional mechanism for fine-tuning tumour suppressor activity, which involves a positive feedback loop where increased protein levels of the tumour suppressor contributes to increased levels of PML via direct transcriptional activation of the PML gene. This fourth mechanisms is involved in the co-regulation of PML protein levels with both p53 and p73/YAP gene expression (de Stanchina et al. 2004; Lapi et al. 2008).

1.2.1.2 PML NBs and the Response to Genotoxic Stress

Mutations arising from an inappropriate response to DNA damage contribute to genome instability and to the development and progression of cancer. As we have discussed above, the PML protein and PML NBs regulate multiple tumour suppressor pathways including p53, which regulates both a cell cycle arrest and apoptosis in response to DNA damage. PML was initial implicated in DNA repair signaling through p53, as PML knock-out mice exhibit an enhanced resistance to radiation induced apoptosis due to attenuated p53 activation (Guo et al. 2000). However, PML NBs have also been implicated in other aspects of DNA recombination and repair (reviewed in [Dellaire and Bazett-Jones 2004]). For example, PML NBs may play a role in the recombination mechanisms responsible for the alternative lengthening of telomeres (ALT) in telomerase-negative tumour cells (Yeager et al. 1999), in sister-chromatid exchange (SCE) via regulation of the Bloom syndrome helicase (BLM) (Zhong et al. 1999), and in the activation of the checkpoint kinase 2 (Chk2) following DNA damage (Yang et al. 2006). Therefore, it is likely that PML may also contribute to tumour suppression by modulating DNA damage signaling and/or repair independently of its role in p53 activation.

One possible mechanism by which PML NBs contribute to DNA damage signaling is through the sequestration/release of DNA repair factors and/or their post-translational modification, some examples of which are listed in Table 1.1 (reviewed in [Dellaire and Bazett-Jones 2004]). One example is the MRE11/RAD50/NBS1 (MRN) complex, which localises to PML NBs in G2 cells prior to DNA damage by ionizing radiation, after which these proteins are released to accumulate at DNA breaks (Lombard and Guarente 2000; Mirzoeva and Petrini 2001). Following the repair of these DNA breaks, the MRN complex relocalises once more to PML NBs (Mirzoeva and Petrini 2001), an interaction that may be mediated by the association of NBS1 with the PML NB component SP100 (Naka et al. 2002). It has been recently shown by Richard and colleagues that the endonuclease activity of MRE11 may become activated by arginine methylation through co-localisation with the protein arginine methyltransferase 1 (PRMT1) at PML NBs (Boisvert et al. 2005). Similarly, localisation of Chk2 kinase at PML NBs can contribute to its activation following DNA damage by promoting its autophosphorylation (Yang et al. 2006).

Another DNA repair kinase that associates with PML NBs is the ataxia-telageictasia mutated and Rad3-related kinase (ATR), which appears to constitutively localise to PML NBs with the replication protein A (RPA) in ALT cells prior to its release in response to DNA damage (Barr et al. 2003). In normal human fibroblasts, ATR does not appear to associate with PML NBs and RPA only associates with PML NBs after DNA damage, where it is found to be juxtaposed to PML NBs after etoposide treatment in cells arrested in S- or G2-phase of the cell cycle (Dellaire et al. 2006a). ATR kinase is recruited to single-stranded breaks after DNA damage by the ATR-interacting protein (ATRIP), which forms a complex with single-stranded DNA coated with RPA (Cortez et al. 2001; Zou and Elledge 2003). The DNA damage response protein TopBP1 also associates with the ATR/ATRIP complex leading to the activation of ATR kinase (Kumagai et al. 2006). TopBP1 protein levels are stabilized by PML expression and, in contrast to the MRN complex, TopBP1 only localises to PML NBs after DNA damage by ionising radiation (Xu et al. 2003). After DNA damage, TopBP1 foci form at PML bodies as soon as 1 h post irradiation and by 8 h these foci also contain Rad50, ataxia-telangiectasia mutated (ATM), breast cancer 1 (Brca1), and BLM (Xu et al. 2003). The possible role in DNA damage signaling played by the accumulation of TopBP1 in PML NBs is unknown. Nonetheless, it is tempting to speculate that its sequestration in these bodies could be regulating ATR kinase activity to promote DNA repair within DNA surrounding the body or to inhibit ATR function in the nucleoplasm.

PML NBs may also serve as sites for the sequestration of unrepaired DNA breaks and/or platforms for DNA damage signaling. For example, PML NBs are often found juxtaposed to focal accumulations of damaged chromatin, known as DNA repair foci, containing the phosphorylated histone variant H2A.X, at late time points following DNA damage (Carbone et al. 2002; Dellaire et al. 2006a; Dellaire et al. 2009). These so-called late DNA repair foci associate with PML NBs at time points much later than described for the accumulation of TopBP1 (i.e. 18 h versus 1–8 h), and are speculated to represent unrepaired DNA breaks that may be essential structures for the amplification of the G1 checkpoint following DNA damage (Yamauchi et al. 2008). In yeast persistent DNA breaks are recruited to another subnuclear domain, the nuclear lamina, where they remain sequestered preventing inappropriate recombination with the rest of the genome (Schober et al. 2009; Oza and Peterson 2010). Yeast lack a PML gene homologue and as a result it is believed that they do not have PML NBs. Therefore, we have speculated that in mammalian cells, rather than sequestration of persistent breaks only at the nuclear lamina breaks may be also sequestered at PML NBs (Dellaire et al. 2009).

Although a direct role for PML NBs in DNA repair has yet to be elucidated, PML NBs can also accumulate single-stranded DNA in ALT cells that is enhanced by UV-irradiation (Boe et al. 2006), and appear to play a role in the replication of DNA tumour virus genomes (Everett 2001). Similarly, PML NBs may also play a role in the replication of mammalian DNA by regulating the localisation and function of BLM during S-phase (Zhong et al. 1999; Eladad et al. 2005). The *BLM* gene is mutated in Bloom syndrome, an autosomal recessive genomic instability syndrome characterized by high levels of sister chromatid-exchange (SCE) (reviewed in [German et al. 1996]) and the BLM protein localises to PML NBs in G1/S of the cell cycle (Zhong et al. 1999). PML null fibroblasts that lack PML NBs exhibit high levels of SCE, presumably due to the mislocalisation or regulation of BLM (Zhong et al. 1999). The localisation of BLM to PML NBs is regulated by its sumoylation, and BLM mutants that cannot be sumoylated induce DNA damage foci containing BLM, Brca1 and γ-H2A.X (Eladad et al. 2005). Together these data support a possible role for PML NBs in preventing genomic instability during S-phase.

Finally, PML NBs can respond to changes in chromatin structure during both S-phase (Dellaire et al. 2006b) and following DNA damage by increasing in number (Mirzoeva and Petrini 2001; Carbone et al. 2002), a process that requires NBS1 and is regulated by ATM, ATR and Chk2 kinase (Dellaire et al. 2006a). As such, changes in the PML NB number may represent an important biomarker for proliferation within tumour biopsies, or genomic instability associated with oncogenic transformation. Equally, PML NB number could also be used to monitor the efficacy of radiation or chemotherapy treatment of cancer by providing a high level read-out of the integrity of DNA damage pathways including ATM, ATR and Chk2 kinase (Dellaire et al. 2006a).

1.2.1.3 PML Controls Cellular Fate by Mediating Apoptosis and Cellular Senescence

In the discussion of the multiple roles of PML and PML NBs in tumour suppression, there is a common thread involving cell fate decisions in response to cellular stress or in development and differentiation. One of the best studied examples of PML NBs controlling cell fate is during programmed cell death or apoptosis (reviewed in [Bernardi et al. 2008; Krieghoff-Henning and Hofmann 2008]). As we discussed with respect to the role of PML in regulating p53 activation following DNA damage, loss of PML impairs the cell's decision to trigger apoptosis in response to genotoxic stress (Guo et al. 2000). In the absence of p53 function, PML can contribute to apoptosis through its interaction and phosphorylation by the Chk2 kinase (Yang et al. 2002). PML is phosphorylated on S117 by Chk2 kinase and mutation of this site to alanine, preventing its phosphorylation, can impair apoptosis in response to ionizing radiation (Yang et al. 2002). PML can also contribute to arsenic trioxide-induced apoptosis by mediating the autophosphorylation and activation of Chk2 (Yang et al. 2006). It is also clear that PML can mediate apoptosis through different pathways depending on the type of genotoxic stress. For example, PML is implicated in UV-induced apoptosis, which is p53-independent and requires the c-Jun-N-terminal kinase (JNK)/c-Jun pathway (reviewed in [Bernardi et al. 2008]). Primary murine fibroblasts that do not express PML are resistant to UV-induced apoptosis and c-Jun transcriptional activity, which is normally activated in response to UV light, is impaired (Salomoni et al. 2005). Although it is unclear how PML regulates c-Jun transcriptional activity, the co-localisation of PML and c-Jun following UV-irradiation (Salomoni et al. 2005) is reminiscent of the interaction with p53 following DNA damage (D'Orazi et al. 2002; Hofmann et al. 2002). Therefore it has been suggested that like p53, c-Jun may be post-translationally modified at PML NBs following UV-irradiation leading to its activation (Bernardi et al. 2008).

In addition to DNA damage-induced apoptosis, loss of PML gene expression also impairs several other extrinsic pathways of apoptosis, including programmed cell death by Fas, tumour necrosis factor alpha (TNF), treatment with ceramide and type I and II interferons (IFNs) (Wang et al. 1998b). For example, apoptosis signaling through the CD95/Fas receptor, involves the PML NB associated protein DAXX (Yang et al. 1997), and PML is required for the proapoptotic functions of DAXX both in response to Fas and mitogenic activation in murine splenocytes (Wang et al. 1998b; Zhong et al. 2000b). Activation of the Fas receptor by its ligand leads to recruitment of the Fas-associated death domain (FADD) protein, which in turn activates both caspase-8 and the JNK pathway (reviewed in [Salomoni and Khelifi 2006]). Fas stimulation causes DAXX to translocate to the cytoplasm where it activates ASK1, which in turn activates JNK leading to apoptosis (Yang et al. 1997; Chang et al. 1998; Charette et al. 2000). In the nucleus, DAXX localisation to PML NBs requires both the SIM domain of DAXX as well as the sumoylation of PML (Ishov et al. 1999; Lin et al. 2006b). Under steady-state conditions, sequestration of DAXX in PML NBs is believed to inhibit the proapoptotic function of DAXX by preventing it from acting as a transcriptional repressor (Li et al. 2000; Lin et al. 2006b). However, many studies suggest both pro- as well as anti-apoptotic functions for DAXX; these studies are described at length elsewhere in two excellent reviews (Salomoni and Khelifi 2006; Krieghoff-Henning and Hofmann 2008).

In addition to DAXX, several other PML NB-associated proteins are implicated in apoptosis, including HIPK2, FLASH, Par-4, THAP1, and ZIPK (reviewed in

[Krieghoff-Henning and Hofmann 2008]) (Table 1.1). As described above, HIPK2 contributes to DNA damage-induced apoptosis by phosphorylating and activating p53 (D'Orazi et al. 2002; Hofmann et al. 2002). However, HIPK2 in cooperation with DAXX has also been implicated in transforming growth factor beta (TGF-β)-mediated apoptosis through the JNK pathway (Hofmann et al. 2003). Flice-associated huge protein (FLASH), also known as caspase 8 associated protein 2 (CASP8AP2), is also involved in Fas-mediated apoptosis (Imai et al. 1999). Upon Fas receptor activation, FLASH translocates from PML NBs in the nucleus to mitochondria where it facilitates the cleavage of caspase-8 and enhances the induction of apoptosis (Milovic-Holm et al. 2007). The THAP domain containing apoptosis associated protein 1 (THAP1) is a proapoptotic factor that recruits the Par-4 protein to PML NBs, and enhance apoptosis in response to serum withdrawal and tumour necrosis factor alpha (TNFα) (Roussigne et al. 2003). Par-4, also known as PRKC, apoptosis, WT1, regulator (PAWR), is a component of PML NBs in vascular endothelial cells (Roussigne et al. 2003). Recently, Par-4 has been shown to be secreted during endoplasmic reticulum (ER) stress in normal and cancer cells, whereby secreted Par-4 induces FADD/caspase-8-dependent apoptosis in cancer cells by binding the glucose-regulated protein-78 (GRP78) (Burikhanov et al. 2009). Over expression of Par-4 can facilitate the association of DAXX and ZIP kinase (ZIPK), which normally associate at PML NBs only in response to IFN-γ or arsenic trioxide (Kawai et al. 2003). These proteins appear to collaborate in the induction of apoptosis, as the overexpression of ZIPK with both Par-4 and DAXX leads to a sixfold increase in apoptosis (Kawai et al. 2003).

The PML NBs also contribute to apoptosis by inhibiting pro-survival pathways, including PI3/PTEN/AKT and NF-κB (Krieghoff-Henning and Hofmann 2008). For example, in response to growth factor stimulation PI3 kinase phosphorylates AKT, which in turn mediates survival in part by translocating to the nucleus where it can phosphorylate the forkhead box (FOXO) transcription factors, triggering their export to the cytoplasm and inhibiting the transcriptional activation of pro-apoptotic genes such as Bim/BCL2-like 11 (BCL2L11) and TRAIL/TNF superfamily, member 10 (TNFSF10) (reviewed in [Tran et al. 2003]). PML can mediate the co-recruitment and dephosphorylation of activated AKT kinase by the phosphatase PP2A, which in turn inhibits AKT's ability phosphorylate and thus repress pro-apoptotic gene expression by the FOXO transcription factors (Trotman et al. 2006). In another example, Tumour necrosis factor alpha (TNFα)-induced apoptosis can be enhanced by overexpression of PML, which inhibits pro-survival signaling through the nuclear factor kappa B (NF-κB) pathway (Wu et al. 2003). NF-κB consists of two subunits (RelA/p65 and p50 or p52) that remain in the cytoplasm as an inactive complex with an inhibitor of NF-κB (IκB) (Shen and Tergaonkar 2009). TNFα induces IκB kinase activation that leads to phosphorylation and subsequent degradation of IκB by proteosome. NF-κB then enters the nucleus and activates the transcription of various target genes, including those encoding the anti-apoptotic protein B-cell CLL/lymphoma 2 (Bcl-2) and the inhibitor of apoptosis proteins (IAPs) 1 and 2 (Shen and Tergaonkar 2009). The PML protein can block this survival pathway by binding the NF-κB subunit RelA/p65, recruiting it to PML NBs, and inhibiting its transcriptional activity (Wu et al. 2003). Finally, specific isoforms of the PML protein may also target pro-survival pathways, providing yet another level of control over cellular fate to apoptotic stimuli. Overexpression of PML IV in the U2OS human osteosarcoma cell line was shown to transcriptionally repress the *Survivin* gene and survivin protein expression is markedly upregulated in *PML –/–* murine embryonic fibroblasts (Xu et al. 2004). *Survivin*, is a member of the IAP gene family that is widely overexpressed in cancer and acts as a regulator of mitosis, as well as playing a broad role as an anti-apoptotic factor during the cellular adaptation to stress (Guha and Altieri 2009). Overexpression of PML IV represses *Survivin* expression and induce apoptosis in both A549 human lung carcinoma and UM-UC-2 human bladder carcinoma cells, suggesting that the modulation of PML IV expression may provide a means of enhancing programmed cell death in a variety of cancers (Xu et al. 2004; Li et al. 2006).

As we discussed earlier in this chapter section, PML and p53 can function together not only in the control of DNA damage-induced apoptosis but also in the induction of cellular senescence in response to both DNA damage and oncogenic Ras (Ferbeyre et al. 2000; Pearson et al. 2000). Cellular senescence, where cells persist in a metabolically active but non-proliferative state, can occur within normal tissues as a result of terminal differentiation, as a consequence of prolonged DNA damage signalling associated with telomere shortening or oncogene expression (Bartkova et al. 2006; Campisi and d'Adda di Fagagna 2007; Mallette and Ferbeyre 2007). As such cellular senescence represents an important mechanism of both aging and tumour suppression. PML overexpression inhibits proliferation and can induce senescence in cancer cell lines, which is associated with increased levels of pRb and cell cycle arrest in G1 (Mu et al. 1997; Le et al. 1998; Ferbeyre et al. 2000). PML can interact with pRb in vivo and pRb has been shown to localise to PML NBs (Alcalay et al. 1998). PML overexpression is associated with the accumulation of hypophosphorylated pRb (Ferbeyre et al. 2000), which in turn leads to the repression of E2F-responsive genes involved in the control the G_1- to S-phase transition (reviewed in [Dyson 1998]). In human fibroblasts, PML-induced senescence could be blocked by expression of the inactivation of pRb by the E7 oncoprotein of the human papilloma virus (HPV) type 16, which suggests that pRb is predominately responsible induction of senescence by PML (Mallette et al. 2004; Bischof et al. 2005). In addition, it was shown that PML isoform IV was predominately responsible for the induction of senescence in fibroblasts; however, *PML –/–* murine embryonic fibroblasts are resistant to PML IV-induced senescence indicating that the cooperation of other PML isoforms may be required for senescence to occur (Bischof et al. 2002). Other cellular changes also occur during the induction of cellular senescence by Ras, including the formation of so-called senescence-induced heterochromatin foci (SAHF), which rely on pRb and coincide with the silencing of E2F-responsive genes (Narita et al. 2003). SAHF contain heterochromatin protein 1 (HP1) and the histone H2A variant macroH2A, and PML NBs appear to play a role in their formation through interactions with both HP1 and the chromatin regulator HIR histone cell cycle regulation defective homolog A (HIRA) prior to the formation of SAHF (Zhang et al. 2005).

Furthermore during the formation of SAHF, HP1 isoform gamma (HP1-γ) is also phosphorylated and this phosphorylation event is required for association of HP1-γ with SAHF but not with PML NBs (Zhang et al. 2007). Thus, similarly to the paradigm of p53 and PML, HP1-γ may be post-translationally modified by phosphorylation at PML NBs prior to deposition in SAHF.

1.2.2 Pathways Regulating PML Protein Stability Represent Possible Therapeutic Targets for Cancer Treatment

The PML protein and PML NBs are implicated in the regulation of a host of tumour suppressor pathways including DNA repair, as we have discussed above (Table 1.1). Therefore, it is not surprising that reduced PML protein levels correlates positively with oncogenic transformation in biopsies from tumours of diverse histological origin including, brain, breast, colon, lung, prostate and stomach (Gurrieri et al. 2004). In addition, it has been shown that proteasomal inhibition can restore PML protein levels in colon and gastric carcinoma cell lines, implicating the active degradation of PML in the etiology of these cancers (Gurrieri et al. 2004). The PML protein is also degraded by gene products expressed by DNA tumour viruses such as the Herpes Simplex Virus 1 (HSV1) and Karposi sarcoma-associated herpesvirus (KSHV) (Everett et al. 1998; Chelbi-Alix and de The 1999; Marcos-Villar et al. 2009). These data raise the question of how PML protein levels are regulated and what role this regulation plays in cancer development and progression? In this section, we will describe the pathways involved in modulating PML protein stability by sumoylation and phosphorylation. Given that PML protein levels are often reduced in cancer, these pathways represent important future avenues for therapeutic intervention, whereby increasing PML protein levels in a tumour could inhibit its growth and/or contribute to enhanced apoptosis following radiation or chemotherapy.

1.2.2.1 Regulation of PML Protein Stability by Sumoylation

A common mechanism identified in the regulation of PML protein stability is the post-translational modification of PML by sumoylation. For example, during infection by HSV1 the viral E3 ligase ICP0 mediates the proteasome degradation of sumoylated species of the PML protein (Everett et al. 1998; Chelbi-Alix and de The 1999). Similarly, during KSHV infection the protein LANA2 enhances the sumoylation of PML by Sumo isoform 2 (Sumo2) followed by its subsequent ubiquitination and proteasomal degradation (Marcos-Villar et al. 2009). Arsenic-induced degradation of PML also requires sumoylation, in particular at K160, and this site is also crucial for LANA2-mediated proteasomal degradation of PML (Lallemand-Breitenbach et al. 2001; Marcos-Villar et al. 2009). The requirement for K160 sumoylation during arsenic-induced PML degradation was not fully understood until the recent discovery that this modification mediates the interaction between PML and the Sumo-dependent E3 ubiquitin ligase Ring finger protein 4 (RNF4) (Lallemand-Breitenbach et al. 2008; Tatham et al. 2008). The PML protein is preferentially modified by polysumo-chains of Sumo2 on K160 and monosumoylation on K490 by Sumo1 (Lallemand-Breitenbach et al. 2008). Treatment of cells with arsenic trioxide results in the monosumoylation of K65 of PML by Sumo1 and the marked accumulation of RNF4 at PML NBs (Lallemand-Breitenbach et al. 2008). RNF4 contains four tandem SIM domains that preferentially bind the polysumo-chains of Sumo2-modified PML at K160, which induces the E3-ubiquitin ligase activity of RNF4 against PML leading to its polyubiquitination and degradation by the proteasome (Lallemand-Breitenbach et al. 2008; Tatham et al. 2008). This RNF4-mediated degradation pathway is also active against the PML-RARα protein, which is also modified on K190 by polysumo-chains of Sumo2 (Lallemand-Breitenbach et al. 2008). Thus, the discovery that RNF4 mediates PML-RARα protein degradation provides a mechanistic explanation for arsenic trioxide induced remission of APL in patients whose promyelocytes carry the t(15;17) translocation (Soignet et al. 1998).

Although the sumo E3 ligase PIAS4 (Piasγ) can localise to the PML NBs (Sachdev et al. 2001; Sun et al. 2005), it remains unclear which E3 ligase(s) is responsible for the sumoylation of PML. Recently, both in vitro and in vivo evidence supports a role for the RAN binding protein 2 (RanBP2) in PML sumoylation (Saitoh et al. 2006; Tatham et al. 2008). These data may explain how mitotic accumulations of desumoylated PML protein can contribute to the formation of new PML NBs containing Sumo1 in early G1 (Dellaire et al. 2006c), which presumably would involve the concomitant sumoylation of PML by RanBP2 during nuclear import. The sumoylation of PML is also reversible through the action of an evolutionarily conserved family of proteases first discovered in yeast to catalyze the removal of sentrin, the yeast homologue of the mammalian Sumo proteins (Li and Hochstrasser 1999). In yeast there are two sentrin-specific proteases (SENPs) known as Ulp1p and Ulp2p, whereas in mammals there are at least six (i.e. SENP1-3 and 5–7) (reviewed in [Mukhopadhyay and Dasso 2007]). There is growing evidence that the SENPs play important roles in tumour suppression, and in particular SENP1 appears regulate apoptosis under different conditions via desumoylation of SIRT1, HIPK1 and PML (reviewed in [Yuan et al. 2008]). For example, over-expression of SENP1 sensitize sinovial fibroblasts derived from patients with rheumatoid arthritis to FAS-mediated apoptosis via PML desumoylation and the release of DAXX from PML NBs (Meinecke et al. 2007). SENP1 appears to be the predominant isoform localised to PML NBs and during HSV1 infection this localisation is enhanced by the viral E3 ligase ICP0, leading to both the desumoylation and degradation of the PML protein (Gong et al. 2000; Bailey and O'Hare 2002). The SENPs also exhibit different specificities for the mammalian Sumo homologues, with SENP3–5 showing preference for Sumo2/3 and SENP1 and 2 showing equal deconjugation activity against all three Sumo paralogues (Mukhopadhyay and Dasso 2007). SENP6 and 7 are of particular interest for PML degradation as they are capable of chain editing and deconjugation of poly-Sumo2/3 chains (Lima and Reverter 2008), which along with Sumo1 chain termination is hypothesized to alter the kinetics of PML degradation by RNF4 (Tatham et al. 2008). However, depletion of SENP1 or SENP6 causes a similar increase in the number and size of PML NBs (Mukhopadhyay et al. 2006; Yates et al. 2008), suggesting a complex relationship exists between Sumo1 and Sumo2/3 deconjugation activities in the maintenance of PML protein levels.

The sumoylation of the PML protein is also stimulated by acetylation of PML. For example, both the over-expression of the acetyltransferase p300 and inhibition of the class I and II HDACs with trichostatin A (TSA) leads to enhanced sumoylation of PML (Hayakawa et al. 2008). However in contrast, depletion of the class II HDAC7 results in the loss of Sumo1-modified species of PML and a reduction in PML NB number (Gao et al. 2008). Although at face value these data seem contradictory, we can rationalize them by the fact HDAC7 can promote PML sumoylation and protein stability independently of its deacetylase activity, most likely through the recruitment of the E2 Sumo ligase, UBC9 (Gao et al. 2008). Similarly, HDAC4 promotes the sumoylation of hypermethylated in cancer 1 (HIC1) independently of its deacetylase activity; however, unlike PML, HIC1 sumoylation is decreased following TSA treatment (Stankovic-Valentin et al. 2007). HDAC inhibitors are being tested in clinical trials, in particular in combination with other chemotherapies, for cancers of varied histological origin and have been approved for the treatment of T-cell lymphoma (Lane and Chabner 2009). Although the mechanism by which acetylation promotes sumoylation of PML remains unclear, the modulation of PML sumoylation and/or protein levels by deacetylase inhibitors may represent an important aspect of the efficacy of these drugs in current clinical trials and should be investigated.

The steady state levels of the PML protein have also been recently demonstrated to be regulated independently of sumoylation at K160 by the ubiquitin E3-ligase known as the human papillomavirus (HPV) E6-associated protein (E6AP) (Louria-Hayon et al. 2003). E6AP contributes to cancer development in HPV infected cells by enhancing the degradation of p53 in cooperation with the HPV E6 protein (Scheffner et al. 1993). Basal PML protein levels are decreased by the over-expression of E6AP, and both the loss of E6AP expression and the expression of a catalytically deficient mutant of E6AP (C833A) can promote increased PML protein stability (Louria-Hayon et al. 2009). Although E6AP functions independently of K160 sumoylation, expression of the C833A mutant of E6AP can block the arsenic trioxide-mediated degradation of PML, suggesting possible cooperation between E6AP and RNF4 in the degradation of PML in response to arsenic (Louria-Hayon et al. 2009).

1.2.2.2 Regulation of PML Protein Stability by Phosphoryation

In addition to sumoylation, PML protein stability is regulated by phosphorylation by at least two kinase pathways, the mitogen-activated protein (MAP) kinase and casein kinase II (CKII) pathways (Hayakawa and Privalsky 2004; Scaglioni et al. 2006). Arsenic trioxide induces the phosphorylation of PML by the MAP kinase extracellular-signal-regulated protein kinase 2 (ERK2), which phosphorylates PML on threonine 28 and serines 36, 38, 40, 527 and 530 (Hayakawa and Privalsky 2004). The phosphorylation of PML by ERK2 is also associated with enhanced PML sumoylation (Hayakawa and Privalsky 2004), which may enhance degradation through binding of RNF4 as discussed above (Lallemand-Breitenbach et al. 2008; Tatham et al. 2008).

The phosphorylation of PML can also regulate its association with the peptidyl-prolyl cis-trans isomerase Pin1, which upon binding targets the PML protein for degradation (Reineke et al. 2008). The binding of Pin1 to PML is mediated by the phosphorylation of four key residues (S403, 505, 518, and 527), one of which is also a target of ERK2 (i.e. 527) (Reineke et al. 2008). These residues surround the SIM domain and the K490 sumoylation site of PML and sumoylation of PML inhibits its interaction with Pin1. Mutation of K490 to arginine (K490R), to prevent the Sumo1-modification of PML at this residue, enhances its degradation following arsenic trioxide (Lallemand-Breitenbach et al. 2008). Given the activation of ERK2 by arsenic trioxide, there is most likely some degree of cross-talk between the effects of phosphorylation and sumoylation of PML on its protein stability, and that Pin1 may be responsible for the enhanced degradation of K490R PML (Petrie and Zelent 2008).

Phosphorylation of PML by CKII can also regulate PML protein stability, where an inverse correlation exists between CKII activity and PML protein levels both in cell lines and in lung cancer samples (Scaglioni et al. 2006). Elevated levels of CKII are found in tumours of every histological origin and CKII is a potent suppressor of apoptosis, promoting survival through phosphorylation of substrates such as IκBα. (reviewed in [Trembley et al. 2009]). The PML protein contains a "degron" sequences between amino acid residues 546–573 that encompasses the SIM domain and is responsible for the poly-ubiquitination of PML and its degradation via the proteasome (Scaglioni et al. 2006). Upon activation by osmotic stress, CKII phosphorylates serines 560–562 and 565 within the degron, and this phosphorylation event triggers the poly-ubiquitination and degradation of PML (Scaglioni et al. 2006). Of the CKII phosphorylation sites, S565 of PML is the most important, as mutation of this residue to alanine drastically reduces both the phosphorylation and CKII-mediated proteasomal degradation of PML. Given the wide-spread activation of CKII in cancer, this pathway represents an attractive therapeutic target by which pharmacological inhibition of CKII would contribute to tumour suppression by enhancing PML protein stability.

1.2.3 Summary of PML NBs and Cancer

PML NBs are critical control centres for cell fate decisions following cellular stresses and DNA damage. They inter-regulate the cell cycle, cell senescence, apoptosis control, and genome maintenance, and these functions place PML NBs at the forefront of tumour suppression within the cell (Dellaire and Bazett-Jones 2004;

Bernardi et al. 2008). They function within these pathways by sequestering and/or facilitating the post-translational modification or degradation of genomic caretakers and cell gatekeepers, including the critical tumour suppressor p53 (Dellaire and Bazett-Jones 2004; Dellaire and Bazett-Jones 2007; Bernardi et al. 2008). Furthermore, they physically respond to alterations in cell proliferation, and to cell stress and DNA damage by changing in biochemical composition and number (Dellaire and Bazett-Jones 2004; Dellaire and Bazett-Jones 2007). As such, they also represent a biomarker for the integrity of stress and DNA damage signaling pathways. The use of PML NBs as a biomarker in cancer is not an entirely new concept. In fact, the presence or absence of PML NBs is used as a biological marker during the diagnosis of APL using immunohistochemistry, and to monitor treatment efficacy, remission status, and the onset of relapse (Dyck et al. 1995). This use of the PML NBs as a marker in APL relies on the fact that the PML-RAR-alpha oncoprotein disrupts the formation of the PML NBs (Dyck et al. 1994; Koken et al. 1994), and treatment of APL cells with all-*trans* retinoic acid results in the reformation of the bodies (Koken et al. 1994). In addition, the now standard treatment of APL with all-*trans* retinoic acid represents one of the first examples of molecularly targeted cancer therapies, a field that is revolutionizing cancer treatments (Sanz et al. 2008; Wang and Chen 2008). Therefore, our increasing knowledge of the biology behind the PML NBs is providing novel insights into carcinogenesis, in additional to the use of these bodies in tumour pathology, as well as new molecular targets for cancer therapy. In particular, the pharmacological manipulation of PML protein levels in solid tumours could prove useful as an adjuvant treatment to enhance cell killing during chemotherapy and radiation treatment.

1.3 The Nucleolus and Cancer

The nucleolus is the largest structure in the nucleus, and is easily visualized using even light microscopy due to its optical density. It was the first subnuclear structure to be observed, and was first noted over 200 years ago by Fontana (Hernandez-Verdun 2006a; Raska et al. 2006). However, it was not until the 1960s that it was identified as the location of ribosomal RNA (rRNA) transcription (Hernandez-Verdun 2006a). As the site of rRNA transcription and ribosome biogenesis, the nucleolus is the canonical example of the relationship between structure and function in the nucleus (Dundr and Misteli 2001; Hernandez-Verdun 2006a). The very existence of the nucleolus depends on on-going rRNA transcription by RNA polymerase I and secondary rRNA processing events (Hernandez-Verdun 2006a; Sirri et al. 2008), and, as will be discussed below, the substructure of the nucleolus is organized around these processes.

Mammalian cells typically contain one to a few nucleoli (Fig. 1.1); however, their size can vary both within a single cell and between cell types. Nucleoli in human cells can range from 0.5 μm in mature lymphocytes to 3–9 μm in proliferating and cancer cells (Hernandez-Verdun 2006a). For example, stimulation of human lymphocyte proliferation by phytohemagglutinin causes a large increase in nucleolar size (Raska et al. 2006). In fact, that association of increased nucleolar size with tumour cells was described as early as 1896 by Pianese (reviewed in [Derenzini et al. 2009]) and with cell proliferation in 1898 by Montgomery (reviewed in [Maggi and Weber 2005]). These observations form the basis of the standing association in the literature between the nucleolus and cancer, both through its function in ribosome biogenesis, and, more recently, through alternative nucleolar functions, including tumour suppression, the cell stress response, cell cycle regulation, cell senescence, and the DNA damage response (Maggi and Weber 2005; Raska et al. 2006; Boisvert et al. 2007; Montanaro et al. 2008; Sirri et al. 2008; Zhang and Lu 2009).

Nucleoli form at transcriptionally active rDNA genes. The rDNA genes are organized in tandem, head-to-tail repeats, termed the Nucleolar Organizer Regions (NORs), and in humans, over 400 copies of the 47-kb repeat are found on the short arms of the acrocentric chromsomes 13, 14, 15, 21, and 22 (Dundr and Misteli 2001; Raska et al. 2006). Each rDNA genes repeat consists of the 18S, 5.8S, and 28S rRNA coding sequences, internal and external transcribed spacers, and an intergenic spacer (reviewed in [Raska et al. 2006]). In mammalian cells, approximately half of the rDNA genes are active, with some NORs being completed repressed, and others containing a mixture of active and repressed genes (Raska et al. 2006). Nucleolar transcription of the rDNA genes is carried out by RNA polymerase I, which is recruited to the rDNA promoters by a preinitiation complex containing transcription initiation factor IA (TIF-IA), the promoter selectivity factor (SL1; also known as TIF-IB in mouse), and upstream binding factor (UBF) (reviewed in [Raska et al. 2006]). TIF-IA, SL1, and UBF are all targets for the positive and/or negative regulation of rRNA transcription during proliferation and cell stress responses, and will be discussed later in these contexts.

Using electron microscopy, three morphologically distinct regions of the nucleolus can be visualized. Fibrillar centres (FCs) are surrounded by the dense fribrillar component (DFC), which are further encircled by the granular component (GC) (Raska et al. 2006; Sirri et al. 2008). Each nucleolus contains approximately 30 FCs, with each FC housing four rDNA genes (Dundr and Misteli 2001). The RNA polymerase I-mediated transcription of the rDNA genes occurs at the interface of the FC and the DFC. The newly-made transcripts radiate out into the DFC, and eventually progress into the GC. Along the way, the precursor (pre-) rRNA is complexed with ribosomal and non-ribosomal proteins into a 90S pre-ribosomal particle. This complex includes small nucleolar ribonucleoproteins (snoRNPs), which contain guide RNAs that specify the locations of the approximately 200 base methylations and pseudouridylations with which the pre-rRNA is modified. The base modifications themselves are carried out by the RNA modifying enzymes, including fibrillarin (a methyl-transferase) and dyskerin (a pseudouridine synthase). The modified pre-rRNA is then cleaved by endonucleases and exonucleases to generate the mature 18S, 5.8S, and 28S rRNAs. Further maturation steps then occur, including addition of the 5S rRNA, which was transcribed in the nucleoplasm by RNA polymerase III, and division of the 90S pre-ribosomal particle into pre-60S and pre-40S subunits.

These pre-subunits are then exported into the cytoplasm, where they finally mature into functional 60S and 40S subunits. The 60S subunit contains the 5S, 5.8S and 28S rRNAs, and approximately 47 RPLs (ribosomal proteins of the large subunit), and the 40S subunit contains the 18S rRNA and approximately 32 RPSs (ribosomal proteins of the small subunit) (reviewed in [Dundr and Misteli 2001; Raska et al. 2006; Dai and Lu 2008; Henras et al. 2008]).

1.3.1 AgNOR Scores as a Prognostic Indicator in Cancer

While the association of nucleolar hypertrophy with cellular proliferation and malignancy has been known for over 100 years, it was not until the identification and standardization of a silver nitrate staining technique (Goodpasture and Bloom 1975; Trere 2000) that the relationship could be properly investigated (Maggi and Weber 2005; Derenzini et al. 2009). This silver staining technique relies on the argyrophilic nature of several nucleolar proteins, specifically the rRNA processing factors nucleophosmin (NPM, B23) and nucleolin, the basal transcription factor UBF, and the largest subunit of RNA polymerase I (Roussel and Hernandez-Verdun 1994). The resulting silver-stained structures are termed the AgNORs, and are visible both in interphase (Maggi and Weber 2005), and at the sites of the rDNA genes in mitosis (Heliot et al. 2000). In interphase, the area of AgNOR staining is related to both nucleolar size and RNA polymerase I activity (Derenzini et al. 2009). Two types of analysis have been used to quantify AgNOR staining: direct counting of the AgNOR dots, and the morphometric method, which uses computer-assisted image analysis to measure the area of AgNOR staining in each cell (Trere 2000). The morphometric method has been proven to be much more objective and reproducible (Trere 2000). Studies that have quantified AgNORs in tumour samples have clearly shown that while the nucleolar parameter cannot be used to diagnose malignancy, it is an important prognostic indicator (Derenzini et al. 2009). Furthermore, because AgNOR size is inversely related to cell doubling time and tumour mass doubling time, it is unique amongst the proliferation markers in that it can be used to gauge tumour proliferation rate, and not just the growth fraction of the tumour (Derenzini 2000). The ability of AgNORs to be used as an indicator of cellular proliferation rate stems from the observation that the faster a cell is progressing through the cell cycle (i.e. the shorter the G1 and G2 periods of the cell cycle are), the faster ribosome biogenesis must be to produce the required amount of protein in a shorter period of time (Pich et al. 2000).

Over 60 studies, representing over 20 different types cancers, have shown that the nucleolar parameter is an independent prognostic variable (reviewed in [Pich et al. 2000; Derenzini et al. 2009]). Cancers analyzed included leukemias, multiple myeloma, melanoma, and carcinomas of the breast, prostate, lung, and colon (reviewed in [Derenzini et al. 2009]). Overall, tumours with high AgNOR scores are poorly differentiated with high metabolic activity, have a abnormal DNA content, and a high proliferation rate, which are all indicative of a malignant phenotype

(Pich et al. 2000). Furthermore, the critical tumour suppressors pRb (Voit et al. 1997) and p53 (Budde and Grummt 1999; Zhai and Comai 2000) can both inhibit rRNA transcription, and human breast cancer tumours with mutated or deleted pRb or p53 have significantly larger nucleoli than tumours with normal pRb and p53 status (Derenzini et al. 2004; Trere et al. 2004). This relationship between tumour suppressor status and nucleolar size provides an additional explanation for the relationship between a tumour's AgNOR score and its malignant potential. Finally, because AgNOR scores can be used to stratify patients into high risk and low risk groups with respect to prognosis, they are therefore useful in identifying patients who might benefit from more aggressive therapy (Pich et al. 2000). However, in the reverse scenario, a low AgNOR score can also be a prognostic indicator of chemotherapy resistance in some cancers, such as has been shown for acute myeloid leukemia (Pich et al. 1998).

Interestingly, another well-established tumour proliferation marker used extensively in pathology, Ki-67 (Scholzen and Gerdes 2000), is actually a nucleolar protein that functions in rRNA transcription (Bullwinkel et al. 2006). In fact, the overexpression of nucleolar proteins as proliferation markers is becoming a common theme in tumour pathology; two other such markers are Nop2/p120 (Saijo et al. 2001) and Mina53 (Zhang et al. 2008). These observations really exemplify the close relationship between the nucleolus, cell growth, and cell proliferation, a topic that will be explored in-depth in the next subsection.

1.3.2 Nucleolar Function, Ribosome Biogenesis and the Inter-Regulation of Cell Growth and Proliferation

The number and size of nucleoli per cell are not constant, and are partly determined by RNA polymerase I activity (Maggi and Weber 2005; Derenzini et al. 2009). Proliferating cells have greater protein requirements than their non-cycling counterparts, and without adequate ribosome biogenesis to fuel cell growth, cells would become progressively smaller after each division (Thomas 2000). Therefore, rRNA transcription is cell-cycle regulated, and there are feedback mechanisms through which nucleolar function regulates the cell cycle to prevent cells from committing to another round of cell division without adequate ribosome biogenesis. The regulation of rRNA transcription is summarized in Table 1.2.

At the most basic level, RNA polymerase I function is regulated during the cell cycle: it is at its highest in S and G2, repressed during mitosis to allow for cell division, and increases throughout G1 (Sirri et al. 2008). During mitosis, the nucleolus disassembles due to RNA polymerase I inhibition through the function of the important cell cycle kinase CDK1 (cyclin dependent kinase 1)-cyclin B complex. CDK1-cyclin B phosphorylates SL1 and TTF-1 (transcription terminator factor-1) (Heix et al. 1998; Sirri et al. 1999). As described above, SL1 is a component of the preinitiation complex for RNA polymerase I, and phosphorylation of SL1 by CDK1 inhibits this function, shutting down RNA polymerase I activity (Heix et al. 1998).

Table 1.2 Positive and negative regulators of rRNA transcription

Positive regulators	Target	Reference
CDK4-cyclin D1, CDK2-cyclin E	UBF	Voit et al. 1999; Voit and Grummt 2001
c-Myc	SL1	Arabi et al. 2005; Grandori et al. 2005
mTOR	SL1	James and Zomerdijk 2004
	TIF-IA	Mayer et al. 2004
MAPK	SL1	James and Zomerdijk 2004
	TIF-1A	Zhao et al. 2003
	UBF	Stefanovsky et al. 2006
PI3K	SL1	James and Zomerdijk 2004
IRS-1, IRS-2	UBF	Tu et al. 2002; Sun et al. 2003; Wu et al. 2005
CKII	TIF-IA	Bierhoff et al. 2008
	UBF	Voit et al. 1992; Lin et al. 2006a
RSK	TIF-1A	Zhao et al. 2003
Negative regulators		
CDK1-cyclin B	SL1	Heix et al. 1998
	TTF-1	Sirri et al. 1999
pRB	UBF	Voit et al. 1997
P130	UBF	Ciarmatori et al. 2001
p53	SL1	Zhai and Comai 2000

The nucleolus only reassembles again in G1 once this repression is relaxed (reviewed in [Boisvert et al. 2007; Sirri et al. 2008]), and rRNA transcription is restarted by the G1 kinases CDK4-cyclin D1 and CDK2-cyclin E in preparation for another potential round of the cell cycle. These kinases phosphorylate UBF, another component of the preinitiation complex, and this promotes its recruitment of RNA polymerase I to the rDNA promoters, activating transcription (Voit et al. 1999; Voit and Grummt 2001).

Because of the increased requirement for ribosomes in rapidly diving cells, growth factors and oncoproteins that promote proliferation, including c-Myc, epidermal growth factor, and insulin-like growth factor, also up-regulate rRNA transcription, increasing the size and number of nucleoli (Raska et al. 2006; Derenzini et al. 2009). c-Myc is an oncogene that functions as a transcription factor for RNA polymerases I, II, and III, and these activities help regulate both cell growth and the cell cycle (Dai and Lu 2008). Its regulation of RNA polymerase I is mediated by its ability to bind rDNA and recruit SL1, ultimately promoting rRNA transcription (Arabi et al. 2005; Grandori et al. 2005). Interestingly, c-Myc is also in a feedback loop with RPL11, a protein of the large ribosomal subunit. *RPL11* is a transcriptional target of c-Myc, but RPL11 protein inhibits c-Myc activity, and when RPL11 is overexpressed, it sequesters c-Myc to the nucleolus, suggesting that this represents one pathway where abnormal nucleolar function could inactivate a positive growth regulator (Dai and Lu 2008). Insulin-like growth factor 1 (IGF-1) has been shown to activation rRNA transcription in numerous ways. First, it and nutrients were shown to activate mTOR (mammalian target of rapamycin), phosphatidylinositol 3-kinase (PI3K), and mitogen-activated protein kinase (MAPK1), which ultimately resulted in greater occupancy of SL1 on rDNA promoters and increased activity of RNA polymerase I (James and Zomerdijk 2004).

Next, insulin receptor substrate-1 (IRS-1) and IRS-2, which are substrates of insulin and IGF-1 receptors, and which translocate to the nucleus following activation, bind to UBF, and upregulate rRNA transcription (Tu et al. 2002; Sun et al. 2003; Wu et al. 2005). Of note, IRS-1 also upregulates rRNA transcription in response to oncogenes such as Simian virus 40 T antigen and v-src (Tu et al. 2002). Next, epidermal growth factor was shown to cause MAPK1 to phosphorylate UBF, increasing the rate of rRNA elongation (Stefanovsky et al. 2006). MAPK1 and RSK (ribosomal protein S6 kinase) kinases have also been shown to phosphorylate TIF-IA, increasing the rate of rRNA transcription (Zhao et al. 2003). The mTOR pathway is frequently hyperactive in cancer, and it promotes cell growth and proliferation, normally responding to signals from nutrient and energy levels (Dowling et al. 2009), including IGF-1 as described above. It is also a target for cancer therapy, and its inhibitor rapamycin causes down-regulation of RNA polymerase I activity through inactivation of TIF-IA, another constituent of the pre-initiation complex (Mayer et al. 2004). Finally, CKII (casein kinase II) also promotes rRNA transcription. CKII is a serine/threonine kinase that is found overexpressed in many cancers, including leukemias and solid tumours (Ruggero and Pandolfi 2003). CKII phosphorylates TIF-IA, triggering the release of RNA polymerase I after initiation and promoting elongation (Bierhoff et al. 2008). It also phosphorylates UBF (Voit et al. 1992), and it recruited to rDNA promoters and functions to stabilize the UBF/SL1 complex, which helps promote the reinitiation of transcription (Lin et al. 2006a).

As a balance to this, and as discussed above, p53 and pRB, which are both negative cell cycle regulators, inhibit rRNA transcription. pRB binds to UBF, inhibiting its ability to bind to rDNA promoters, and causing a reduction in rRNA transcription (Voit et al. 1997). Another pRB family member, p130, is also able to regulate UBF in a similar manner (Ciarmatori et al. 2001). p53 can physically bind to SL1, preventing its interaction with UBF, and inhibiting rRNA transcription (Zhai and Comai 2000). Next, the nucleolar protein and tumour suppressor ARF, which will be discussed in much more depth in the next subsection in relation to its role in p53 regulation, has been shown to inhibit the processing of the pre-rRNA transcripts (Sugimoto et al. 2003).

Finally, to ensure that cells do not divide without sufficient growth, normal ribosome biogenesis is required for the expression of cyclin E (Volarevic et al. 2000). Cyclin E is a key G1 regulator of CDK2 (Kaldis and Aleem 2005), and without active CDK2/cyclin E, pRb remains unphosphorylated, and bound to and inhibiting the E2F transcription factor. E2F target gene products are required for the cell to progress through the restriction point, the point at which the cell commits to initiating DNA replication and completing the cell cycle (Harbour and Dean 2000; DeGregori 2002; Genovese et al. 2006).

1.3.3 Nucleolar Functions Beyond Ribosome Production

It has become clear that the role of the nucleolus in cells extends far beyond ribosome biogenesis. The advent of advanced microscopy techniques, including fluorescence microscopy and live-cell imaging (Dellaire et al. 2003; Mayer and Grummt 2005;

Olson and Dundr 2005; Politz et al. 2005; Handwerger and Gall 2006; Hernandez-Verdun 2006a; Hernandez-Verdun 2006b; Trinkle-Mulcahy and Lamond 2007) and large-scale proteomic studies using mass spectrometry (Andersen et al. 2002; Scherl et al. 2002) have allowed much more extensive analyses of nucleolar dynamics, structure, and biochemical composition. For example, of the over 700 human proteins identified in the nucleolus, only approximately 30% function in ribosome biogenesis (Boisvert et al. 2007). Other nucleolar proteins have functions that include the processing and maturation of non-nucleolar RNA and ribonucleoproteins, messenger RNA (mRNA) export, and the regulation of the cell cycle, DNA replication, DNA repair, cell senescence, tumour suppression, and the cell stress response (Maggi and Weber 2005; Raska et al. 2006; Boisvert et al. 2007; Dellaire and Bazett-Jones 2007; Montanaro et al. 2008; Sirri et al. 2008).

1.3.3.1 The Nucleolus, p53, ARF, and the Cell Stress Response

Over the last 10 years, the nucleolus has emerged as a critical regulator of the p53 tumour suppressor. As described in the section on PML NBs, p53 is a cell gatekeeper that regulates the cell cycle checkpoints, apoptosis (programmed cell death), and cell senescence (Kinzler and Vogelstein 1997; Levine 1997; Rodier et al. 2007). Over 50% of cancers contain p53 mutations (Soussi and Lozano 2005), highlighting the importance of p53 as a cellular guardian. Normally, p53 proteins are maintained at very low levels through the actions of its negative regulator: MDM2 (mouse double minute 2), an E3 ubiquitin ligase that targets p53 for degradation (Iwakuma and Lozano 2003; Moll and Petrenko 2003). p53 is stabilized following cellular stress (Rubbi and Milner 2003; Olson 2004; Mayer and Grummt 2005), DNA damage (Rubbi and Milner 2003; Gjerset 2006), and aberrant oncogene expression (Weber et al. 1999; Saporita et al. 2007), allowing it to transactivate its transcriptional targets. These transcriptional targets can include the pro-apoptotic genes such as Noxa, Puma, Bid, and Bax, and the critical G1/S regulator p21 (Menendez et al. 2009). One of the primary methods of p53 stabilization is the interruption of the MDM2-p53 interaction. Nucleolar proteins are key in this process (Rubbi and Milner 2003), with the nucleolar tumour suppressor ARF (Alternate Reading Frame) being the prime example (Saporita et al. 2007). Therefore, through its role in p53 regulation, the nucleolus also regulates the cell cycle checkpoints, apoptosis, and cell senescence.

ARF is the result of the alternate reading frame of the *INK4a/Arf* locus. This locus produces both the p16^{INK4a} and ARF proteins, whose different mRNA transcripts are the result of a frameshift induced by alternate splicing. Both p16^{INK4a} and ARF function independently as tumour suppressors, and one of ARF's most important roles is as a regulator of p53 through its interaction with MDM2. ARF is normally found predominantly in the nucleolus; however, when nucleolar structure is disrupted, for example, due to RNA polymerase I inhibition, ARF translocates to the nucleoplasm where it binds MDM2, preventing its interaction with p53 (Saporita et al. 2007). In a similar scenario, following DNA damage by ionizing radiation or Mitomycin C, or when ARF is upregulated in response to oncogenic signals, including those from Ras and c-Myc, ARF can sequester MDM2 in the nucleolus, again preventing its interaction with p53 (Weber et al. 1999; Khan et al. 2004; Dias et al. 2006; Saporita et al. 2007). ARF targets MDM2 to the nucleolus via its own nucleolar localization signal, and also through a cryptic nucleolar localization signal on MDM2 that is exposed following a conformational change in MDM2 induced by ARF binding (Saporita et al. 2007).

Other nucleolar proteins that can regulate p53 include nucleostemin, NPM, nucleolin, and the ribosomal proteins RPL5, RPL11 RPL23, and RPS7 (Mayer and Grummt 2005; Takagi et al. 2005; Gjerset 2006; Saxena et al. 2006; Derenzini et al. 2009; Zhang and Lu 2009), through interactions that can involve p53, MDM2, ARF, and each other. Therefore, the model that has been developed is that the nucleolus represents a key stress sensor. Cellular stresses that cause RNA polymerase I inhibition induce nucleolar disruption, initiating a release of proteins that can modulate the function of p53, activating cell cycle checkpoint or apoptotic pathways (Rubbi and Milner 2003; Olson 2004; Mayer and Grummt 2005; Gjerset 2006). The list of stresses known to inhibit RNA polymerase I includes hypoxia, heat shock, transcriptional inhibitors, topoisomerase I inhibitors, and some DNA damaging agents (e.g. UV-irradiation and cisplatin) (Rubbi and Milner 2003). For stresses that do not directly inhibit rRNA transcription, kinases, such as JNK2 (c-jun N-terminal protein kinase 2), function to halt transcription (Mayer et al. 2005; Mayer and Grummt 2005). The importance of the nucleolus to p53 activation is further supported by the observation that when there is a high level of DNA damage that does not affect the nucleolus (i.e. when UV-induced damage is specifically targeted to a non-nucleolar site), p53 is not activated (Rubbi and Milner 2003).

1.3.3.2 Nucleostemin

This system of regulating p53 results depends on the dynamic movement of many nucleolar proteins, and there appears to be a delicate relationship between their relative levels. For example, nucleostemin, which is a marker for many stem cells and cancer cells, can suppress p53 after nucleolar disassembly by interfering with the interaction between MDM2 and RPL23 (Meng et al. 2008). In a contradictory scenario, ectopic nucleostemin has been shown to also activate p53 by binding to and inhibiting MDM2 function (Dai et al. 2008). Furthermore, depletion of nucleostemin causes increased binding of RPL5 and RPL11 to MDM2, again activating p53 (Dai et al. 2008). Therefore, the level of nucleostemin is closely monitored by the p53 pathway, with both depletion and overexpression of nucleostemin leading to p53 stabilization (Ma and Pederson 2008). It can also apparently function as both an oncoprotein and a tumour suppressor, with the ultimate outcome potentially depending on the cellular context (Dai et al. 2008; Meng et al. 2008). Recently, nucleostemin has also been shown to be required for nucleolar structure, rRNA processing, and telomerase activity (Romanova et al. 2009a; Romanova et al. 2009b), providing additional explanations for why it is expressed in stem and cancer cells, and why depletion would induce a p53-dependent arrest.

1.3.3.3 Nucleophosmin

Like nucleostemin, NPM is also implicated in cancer development, especially haematological malignancies. For example, approximately 35% of adult acute myeloid leukemia patients have mutations in NPM that cause it to be relocalized to the cytoplasm, and these mutations are thought to be tumour-initiating (Gjerset 2006). Furthermore, the nucleophosmin gene, *NPM1*, is fused to the anaplastic lymphoma kinase (*ALK*) gene in the most frequent translocation found in anaplastic large cell lymphoma, the myeloid leukemia factor 1 (*MLF1*) gene in myelodysplastic syndrome, and retinoic acid receptor alpha (*RARA*) gene in acute promyelocytic leukemia (APL) (Naoe et al. 2006). However, like nucleostemin, NPM can have both tumour suppressive and oncogenic functions. NPM, which functions in rRNA processing and ribosomal biogenesis, also functions to stabilize and sequester ARF in the nucleolus (Gjerset 2006). NPM is also required for the proper compartmentalization of the nucleolus, and this function is regulated through phosphorylation by CKII (Louvet et al. 2006). However, following cellular stress, NPM can also interact with MDM2, leading to the stabilization of p53 (Kurki et al. 2004). NPM can also directly interact with p53, modulating its transcriptional activities (Colombo et al. 2002; Maiguel et al. 2004). The observations that NPM can both activate and suppress p53 function suggests that the relative levels of NPM, ARF and MDM2 are critical in determining the ultimate outcome on p53 activity (Gjerset 2006). The oncogenic and growth promoting functions of NPM are also linked to both its function in ribosome biogenesis (Gjerset 2006) and its ability to enhance c-Myc function (Li et al. 2008). NPM also functions in genome maintenance, as it is involved in the regulation of centrosome duplication (Gjerset 2006).

1.3.3.4 Nucleolin

Nucleolin is also associated with cancer development, as it is highly expressed in rapidly proliferating cells, and can also be found on the cell surface in a wide variety of cancer cells (Storck et al. 2007). However, like nucleostemin and NPM, it has both tumour suppressive and oncogenic functions. Nucleolin functions in both RNA polymerase I and II transcription, as well as rRNA processing, and mRNA stabilization (Storck et al. 2007). Its role in RNA polymerase I transcription is related to its ability to bind and stabilize the G-quadruplex structures in the rDNA, which increases the density of RNA polymerase I loading, upregulating rRNA transcription. The drug CX-3543 is an inhibitor of the nucleolin/rDNA complex, and it is currently in human clinical trials for carcinoid/neuroendocrine tumours (Drygin et al. 2009). Nucleolin may also contribute to oncogenesis through telomere maintenance, as it can bind both telomere repeats and the catalytic component of telomerase (Storck et al. 2007). However, nucleolin also responds to stress to help inhibit DNA synthesis and the cell cycle. First, following heat shock, nucleolin relocalizes to the nucleoplasm, binding to replication protein A and inhibiting DNA replication (Daniely and Borowiec 2000). Next, like the other nucleolar proteins, nucleolin can bind to and prevent MDM2 from ubiquitinating p53, ultimately activating p53 (Saxena et al. 2006). Nucleolin can also help regulate p53 at the level of translation. After DNA damage, nucleolin and RPL26 can bind to the 5' untranslated region of p53 mRNA, increasing the rate of p53 translation (Takagi et al. 2005). Finally, nucleolin can also bind p53 directly following relocalization to the nucleoplasm induced by ionizing radiation or camptothecin treatment, and this interaction is required for nucleolin to relocalize (Daniely et al. 2002).

1.3.3.5 Other Tumour Suppressors

Other tumour suppressors known to localize to the nucleolus include ING1 (inhibitor of growth 1) and PML. ING1 can interact with ARF, and it also relocalizes to the nucleolus following UV-irradiation, facilitating apoptosis (Scott et al. 2001; Gonzalez et al. 2006; Zhu et al. 2009). The PML protein is the main component of nuclear structures known as PML nuclear bodies (Dyck et al. 1994; Koken et al. 1994). PML NBs are thought to function in gene transcription, proteasomal degradation, the viral response, tumour suppression, apoptosis, cell senescence, and DNA repair (see Section 2.2 and [Dellaire and Bazett-Jones 2004; Bernardi and Pandolfi 2007]). Following treatment with doxorubicin (a DNA intercalator and topoisomerase II inhibitor), PML can sequester MDM2 in the nucleolus via an interaction with the ribosomal protein L11, which results in the stabilization of p53 (Bernardi et al. 2004).

1.3.3.6 The Nucleolus and Genome Maintenance

As discussed above, the nucleolar protein NPM functions in genome maintenance through its regulation of centrosome duplication (Gjerset 2006). However, there are additional mediators of genome maintenance found in the nucleolus. The most well-known example is telomerase. The telomerase enzyme is a ribonucleoprotein complex that extends the telomeres, the terminal segments of the chromosomes, which would otherwise become successively shorter each DNA replication (Collins 2006). In the majority of human cancers, telomerase is upregulated, which allows the cells to replicate indefinitely (Tian et al. 2009). However, telomere dysregulation can cause genomic instability, which may also lead to oncogenesis (Bailey and Murnane 2006). Active telomerase requires the telomerase reverse transcriptase (TERT), the telomerase RNA component (TERC), and the nucleolar ribosomal pseudouridine synthase and RNA-binding protein dyskerin (Cohen et al. 2007; Kirwan and Dokal 2009). The association of telomerase with the nucleolus appears to be for both the RNP's processing and maturation, as well as for its regulation (Raska et al. 2006; Derenzini et al. 2009). For example, under normal conditions, telomerase is sequestered in the nucleolus until S phase; however, in malignant cells, it is dispersed into the nucleoplasm, indicating that this level of regulation has been lost (Wong et al. 2002). Of note, ionizing radiation induces telomerase to relocalize back to the nucleolus in both primary and tumour cells, ostensibly to prevent telomerase from aberrantly affecting DNA double-strand break ends (Wong et al. 2002). Because senescence and telomere shortening are intimately linked, telomerase regulation by the nucleolus represents another example of how the nucleolus can function in the control of cell senescence (Johnson et al. 1998).

The nucleolar protein Bop1 (block of proliferation 1) and its associated complex are involved in rRNA processing and ribosome biogenesis; however, both over expression and depletion of Bop1 lead to abnormal mitoses, identifying a function for this protein in maintaining chromosomal stability (Strezoska et al. 2000; Killian et al. 2004). Furthermore, expression of a dominant-negative version of Bop1 leads to p53 activation, suggesting that it might also function in the nucleolar stress response (Maggi and Weber 2005). Recently, 18S rRNA has been shown to help promote genomic stability by being part of an RNA/protein complex with Myc-induced SUN domain-containing protein (Misu; a nucleolar RNA methyltransferase) that translocates to and is required for proper assembly of the mitotic spindle (Hussain et al. 2009). Other proteins involved in genome maintenance and/or DNA repair with known nucleolar associations include the telomere repeat binding factor TRF2 (Zhang et al. 2004); the DNA helicases RecQL4 (Woo et al. 2006), BLM (Sanz et al. 2000; Yankiwski et al. 2000), and WRN (Marciniak et al. 1998); the endonuclease Mus81 (Gao et al. 2003); Rad17 (Chang et al. 1999); Rad50 (Andersen et al. 2002); Rad52 (Liu et al. 1999); hMSH6 (Mastrocola and Heinen 2009), and the DNA damage response kinase ATR (ATM- and Rad3-related) (Andersen et al. 2002).

Some of these proteins may localize to the nucleolus to help maintain the stability of the rDNA repeats, especially during S phase, including BLM, Mus81, and Rad52. BLM is a RecQ helicase, and mutations in its gene cause Bloom syndrome, a human genomic instability syndrome whose characteristics include increased sister-chromatid exchange and cancer predisposition. The BLM helicase has been shown to act on the rDNA repeats during replication, as it specifically binds to the non-transcribed spacer region of the rDNA repeat, which is where replication forks initiate. Therefore, BLM may help maintain the stability of this repetitive region (Schawalder et al. 2003). Of note, RecQL4 and WRN are in the same helicase family as BLM, and mutations in their genes cause Rothmund-Thomson syndrome and Werner syndrome, respectively. Like Bloom syndrome, these are human genomic instability syndromes characterized by cancer predisposition (Hanada and Hickson 2007). Therefore, these helicases may be functioning in a similar manner to BLM in the nucleolus. However, WRN translocates out of the nucleolus in response to ionizing radiation, but not UV-irradiation, suggesting that nucleolar localization of this protein may also be functioning to regulate its activity (Karmakar and Bohr 2005). Next, Mus81 is an endonuclease whose targets include Holliday junctions and replication forks. It is most abundant in S phase, and hyper-accumulates in the nucleolus, suggesting that it too might be required for rDNA replication (Gao et al. 2003). Finally, Rad52, which functions in homologous recombination repair, also specifically localizes to nucleoli during S phase (Liu et al. 1999), indicating that it too may function in rDNA replication.

Other genome maintenance proteins might be sequestered in the nucleolus to prevent them from indiscriminately eliciting a DNA damage signal and functioning on DNA inappropriately. Examples of this include telomerase and WRN, which were discussed above, and TRF2, Rad17, and hMSH6. TRF2, which besides stabilizing telomere ends, functions in the DNA double-strand break response (Bradshaw et al. 2005), and also stimulates the activity of BLM and WRN (Opresko et al. 2002). It is present in the nucleolus early in the cell cycle, but begins to leave during G2. However, it begins to return to the reforming nucleoli during cytokinesis. Therefore, besides its role in telomerase regulation, the nucleolus may further regulate telomere function by controlling the activity of TRF2 (Zhang et al. 2004). Next, Rad17 is a replication factor C-like factor that loads the 911 (Rad9, Rad1, and Hus1) complex onto a DNA template. The 911 complex is structurally similar to the PCNA sliding clamp that is required for DNA replication. Loading of the 911 complex is required for the ATR-mediated activation of the checkpoint kinase Chk1, and may help to localize proteins required for the DNA damage response (Shechter et al. 2004). Normally, Rad17 is sequestered in the nucleolus, but it is released following UV-irradiation, suggesting that the nucleolus may regulate Rad17 function by preventing it from interacting with chromatin in the absence of DNA damage (Chang et al. 1999). Finally, hMSH6, which functions in mismatch repair, is hyper represented in the nucleolus compared to in the nucleoplasm. Following DNA damage by an alkylating agent, the nucleolar fraction of hMSH6 was reduced. Therefore, the nucleolar localization of hMSH6 may help prevent the mismatch repair system from functioning in the absence of DNA damage (Mastrocola and Heinen 2009).

Rad50, which is a constituent of the damage sensing MRN complex, and the critical DNA damage response kinase ATR (which responds to stalled replication forks and UV-induced DNA damage) were found associated with the nucleolus in a proteomics screen (Andersen et al. 2002), so the significance of these associations is unknown.

1.3.4 Nucleolar Function as a Cancer Initiator

While it is clear that the increased ribosome biogenesis by the nucleolus frequently occurs in cancer cells to help fuel their proliferation, one of the questions being discussed in the literature is whether or not altered nucleolar function, specifically altered ribosome biogenesis, could in fact be a cancer-initiating event. There are several inherited human disorders that support this hypothesis. Diamond-Blackfan anemia (DBA), which is an anemia characterized by a specific decrease in erythroid precursors, is associated with heterozygous mutations in one of several ribosomal protein genes: RPS7, RPS17, RPS19, RPS24, RPL5, RPL11, and RPL35a (Badhai et al. 2009; Zhang and Lu 2009). The disorder is also associated with cancer predisposition, suggesting that the alterations in ribosome biogensis initiated by these RP gene mutations are oncogenic (Badhai et al. 2009; Zhang and Lu 2009). However, it is important to note that some of these proteins have known non-ribosomal function. For example, RPS7, RPL5, and RPL11 all bind MDM2, inhibiting its negative regulation of p53 (Badhai et al. 2009; Zhang and Lu 2009).

Next, dyskeratosis congenital (DC), which is a rare X-linked disorder that is characterized by progressive bone marrow failure and that includes a predisposition to cancer, is caused by mutations in the *DKC1* gene. The protein product of this gene is dyskerin, which functions in snoRNPs and telomerase as a pseudouridine synthase (Ruggero and Pandolfi 2003; Montanaro et al. 2008). While its contribution to carcinogenesis as a component of the telomerase complex cannot be discounted, there are data that suggest that the cancer predisposition is at least partially due to abnormal ribosome biogenesis. For example, *DKC1* hypomorphoic mice develop tumours before their telomeres are shortened enough to cause genomic instability (Ruggero et al. 2003). Furthermore, DC patients and the DKC1 hypomorphic mice exhibit a selective defect in translation of a group of mRNAs that contain internal ribosome entry sites, such as the mRNA for the p27 tumour suppressor (Yoon et al. 2006). Therefore, it is possible that the altered translation caused by mutant dyskerin may help induce carcinogenesis (reviewed in [Ruggero and Pandolfi 2003; Montanaro et al. 2008]).

Two other genetic disorders that are caused by mutations affecting ribosome biogenesis and that are associated with cancer predisposition are cartilage hair hypoplasia and Shwachman-Diamond syndrome. Cartilage hair hypoplasia is caused by mutations in the non-coding RNA component of the ribonucleoprotein complex RNase MRP (RMRP). RMRP functions in ribosomal RNA processing, specifically in processing of the 5.8S rRNA (reviewed in [Montanaro et al. 2008]). Shwachman-Diamond syndrome is caused by mutations in the SBDS gene, whose protein product functions in rRNA processing and ribosome maturation (reviewed in [Montanaro et al. 2008]). Therefore, altered translation may be able to induce carcinogenesis in much the same way as altered transcription, with both quantitative and qualitative changes in translation potentially being able to upregulate oncoproteins and/or downregulate tumour suppressors. Therefore, because it is the birthplace of ribosomes, the nucleolus could function in tumorigenesis by modifying the ribosomes it produces, ultimately affecting how and what they translate. However, the non-ribosomal functions of these proteins clearly cannot be discounted (reviewed in [Ruggero and Pandolfi 2003; Montanaro et al. 2008]).

1.3.5 Summary of the Nucleolus and Cancer

Our understanding of the nucleolus as a dynamic and "plurifunctional" structure (Pederson 1998) has ushered in a new era of being able to functionally analyze the nucleolus. As such, the nucleolus has emerged as a critical stress sensor, and as a regulator of the cell cycle and genome stability. How and why the nucleolus is involved in these seemingly disconnected pathways are extremely active areas of research that are providing insights into the development of cancer, and we now have the tools, in the form of advanced imaging techniques and proteomic screens, to evaluate nucleolar dynamics in ways that were not possible 20 years ago.

However, we can begin to speculate about the how and why of non-ribosomal nucleolar function through the very structure and function of the nucleolus. The nucleolus and other subnuclear structures, including the PML NBs, represent protein islands within an ocean of chromatin (Dundr and Misteli 2001; Politz et al. 2005). Therefore, they provide unique cellular environments that are separated from the majority of mRNA transcription, providing a site for protein regulation through sequestration. In fact, the ability of nucleolar proteins to regulate p53 stems from this very attribute of the nucleolus. ARF, nucleostemin, and NPM do not normally come into large-scale contact with MDM2 and p53; therefore, their release following nucleolar disassembly allows these proteins to interact specifically under stress conditions (Rubbi and Milner 2003; Olson 2004; Mayer and Grummt 2005; Gjerset 2006).

The nucleolus also contains a wide-array of RNA modifying and processing enzymes, which might explain why ribonucleoproteins such as telomerase and signal recognition particle, and non-ribosomal RNAs such as tRNAs are partially processed and matured here (Raska et al. 2006). Next, ribosome biogenesis is the major metabolic activity of the cell (Schmidt 1999), with 14, 000 ribosomal subunits leaving the nucleolus per minute, and approximately 60% of the total RNA transcription in the cell being rRNA (Raska et al. 2006). Therefore, we can speculate that even minor perturbations in the cellular environment, not to mention significant events like DNA damage, would affect rRNA transcription before other nuclear functions. It would then make sense for the nucleolus to function as a front-line sensor of the cellular environment. Finally, the very fact that cell growth (which is regulated by the level of ribosome biogenesis) and cell proliferation must be co-regulated means that the nucleolus is already involved in the feedback mechanisms that integrate these processes (Thomas 2000; Volarevic et al. 2000). In conclusion, nucleolar number, morphology, and biochemical composition reflect underlying cellular conditions, such as proliferation rate, tumour suppressor and oncogene status, stress, and DNA damage, and these parameters reflect and can be used to assess the overall state of the cell.

1.4 The Perinucleolar Compartment in Cancer Cells

During carcinogenic transformation, cellular structure and function undergo significant alterations. Changes in gene expression and function are often reflected in structural changes in cancer cells. Notably, the morphological characteristics of cancer cell nuclei are drastically different from normal ones and the level of alteration is directly associated with the progression of the diseases, such that the nuclear morphometry is one of the key criteria in tumour grading clinically. The perinucleolar compartment (PNC), a subnuclear structure at the periphery of the nucleolus, is one of the nuclear structural changes that take place during malignant transformation. In this section, we will discuss the current understanding of the structure and functional implication of the PNC in malignant cells.

The PNC was first shown in the isolation and identification of the polypyrimidine track binding protein (PTB), in which immunolabeling using anti-PTB antibody showed a perinucleolar enrichment in HeLa cells (Ghetti et al. 1992). However, the structure is not readily detectable through conventional histological stains. When cells are examined by a transmission electron microscope with standard labeling, the PNC appears to be composed of dense reticulated thick strands in direct contact with the nucleolus, but is structurally distinct from nucleolus. It is irregular in shape and the size ranges from less than 1–5 um (Huang et al. 1997) (Fig. 1.1).

1.4.1 The PNC Formation is Associated with Malignant Phenotype In Vitro and In Vivo

Earlier studies showed that PNC prevalence (the percentage of cells containing at least one PNC) is significantly increased in cancer cell lines in vitro (Huang et al. 1997). A more recent survey of a large number of normal and malignant cells either transformed in vitro or isolated from human tumours further supports the association of PNCs with cancer cells (Norton et al. 2008a). PNCs are absent in normal primary cells including mouse and human embryonic stem cells and are rare in immortalized cells. In contrast, PNC prevalence is increased and is highly heterogeneous in different cancer cell lines derived from solid tissue origin, with some reaching as high as nearly 100%. To determine whether the PNC prevalence within a given cell population is associated with the level of malignancy, Norton et al, examined a series of prostate cancer cell lines (PC3) derived from the same tumour, but representing varying levels of metastatic capability. The results show that PNC prevalence closely correlates with the metastatic capacity of the cell lines, indicating that a high PNC prevalence is associated with a strong capacity for metastasis (Norton et al. 2008a).

To evaluate the association of PNCs with cancer in vivo, Kamath et al. investigated PNC prevalence in human breast cancer samples. Double blinded studies showed that PNC prevalence increases in parallel with disease progression as characterized by clinical stages and histological grades. PNC prevalence shows a stepwise elevation from the primary disease to lymph nodes and to metastasis, reaching nearly 100% in distant metastases, demonstrating that PNC prevalence positively correlates with the disease progression of cancer (Kamath et al. 2005). Additionally, multivariate studies using case-matched samples showed that PNC prevalence has independent prognostic information over tumour size and grading in stage 1 breast cancer patients. The findings both in vitro and in vivo demonstrate that PNC formation is closely associated with malignancy. The enrichment of PNC containing cells to nearly 100% in distant metastases suggests that PNC formation reflects key changes during transformation that confer metastatic capabilities and that PNC containing cells have growth and metastatic advantages over cells without PNC. Thus, PNC could serve as a pan-cancer marker for tumours of solid tissue origin (Norton et al. 2008a).

1.4.2 Association of PNC with Pol III Transcription

The PNC is a multi-component structure enriched with RNA binding proteins and RNAs. The RNA binding proteins so far identified in the PNC include polypyrimidine tract binding protein (PTB), CUG-BP (Huang et al. 1998; Huttelmaier et al. 2001), KSRP (Hall et al. 2004), Raver1 (Huttelmaier et al. 2001), Raver2, Rod1, and nucleolin (Pollock and Huang 2009). With the exception of nucleolin, these proteins have primarily been implicated in pre-mRNA metabolism. PTB participates in a broad range of functions in post-transcriptional processing of pre-mRNA. It plays an important role in alternative splicing, RNA trafficking, RNA stability, and translational controls (Sawicka et al. 2008). KSRP and CUB-BP are both also involved in conventional and alternative splicing, and regulation of the process (Castle et al. 2008). More recently, KSRP has been shown to be involved in microRNA biogenesis (Trabucchi et al. 2009). Raver 1 and 2 are PTB binding proteins and mediate splicing repression (Rideau et al. 2006). Nucleolin is a multifunction protein that has been implicated in pre-rRNA processing and ribosome assembly, as well as in pre-mRNA metabolism and cell surface functions (Mongelard and Bouvet 2007). Although many pol I transcription and pre-rRNA processing factors, including UBF1, RNA polymerase I, TAF1, fibrillarin, and NPM, have been examined for their association with the PNC, nucleolin is the only nucleolar protein found enriched in the PNC (Chen and Huang unpublished data). The PNC associated RNAs include RNase MRP, RNAse P, hY, SRP, and Alu RNAs (Matera et al. 1995; Lee et al. 1996; Wang et al. 2003). These RNAs are co-enriched in the PNC with the RNA binding proteins, in which they are detected above the nucleoplasmic level by immunofluorescence. RNase MRP and P are ribozymes involved pre-ribosomal RNA and tRNA processing (Xiao et al. 2002. hY and Alu RNA form RNPs with proteins and function of these RNPs are still at the beginning of investigation. Interestingly, ribosomal RNAs (Xiao et al. 2002) are not found enriched in the PNC (Matera et al. 1995).

Several lines of evidence indicate that the PNC is associated newly synthesized RNA. A 5-min pulse labeling with BrU shows that the PNC is highly concentrated with newly synthesized RNA (Huang et al. 1998). Selective inhibitions of individual polymerases demonstrate that these RNA are mainly synthesized by RNA polymerase III. Specific pol III inhibitor treatment induces a disassembly of the PNC within 2–3 h, suggesting that the PNC is dependent upon a continuous synthesis of pol III transcripts (Wang et al. 2003). Together with the evidence of co-localization of the RNA and RNA binding protein, these observations lead to a working hypothesis that the PNC is enriched with novel RNA-protein complexes that are involved in the metabolism of newly synthesized pol III RNA. It appears that not all pol III

RNAs can be found concentrated in the PNC since in situ hybridization of other pol III RNA including 5S RNA, U6 RNA and tRNAs fails to detect enrichment of these RNAs in the PNC (Matera et al. 1995).

1.4.3 The PNC is Nucleated upon DNA

In addition to the requirement for ongoing pol III transcription, PNC structural maintenance also requires the integrity of DNA. Treatment of cells with DNA damaging agents or UV irradiation induces PNC disassembly. Blocking the DNA damage repair response does not influence the disassembly of PNC, but it does block the recovery of PNC prevalence post-DNA damage. These findings indicate that the DNA damage itself and not the repair is responsible for the disruption of the PNCs (Norton et al. 2008b). Examination of a large panel of DNA damaging agents reveals that not all types of DNA damage disrupts PNCs. Only those capable of DNA intercalations or cross-linking disassemble the PNC. For some, the treatment disrupts PNC without significant interruptions of pol III transcription. The uncoupling between pol III transcription and PNC disassembly demonstrates that DNA structural integrity is critical to the PNC structure. The association of the PNC with the DNA is further supported by several cell biological observations. (1) The PNC colocalizes with BrU labeled nascent RNA as detected by confocal microscopy. (2) The number of PNCs within each nucleus increases in parallel with the increased number of endo-replication cycles in temperature sensitive cdk1 mutant cells at the non-permissive condition. These cells continue to replicate DNA, but are unable to divide at the non-permissive condition (Laronne et al. 2003). The increase in the number of PNC corresponds to the number of rounds of DNA duplication. (3) PNCs appear to split into two dots during S phase, similar to the behavior of DNA replication as observed by in situ hybridizations, and return to one structure in G2; (4) Releasing cells into S phase from a hydroxyurea block at G1/S demonstrated that most of the split PNCs are observed at early to mid S phase, suggesting an association with a more gene rich chromatin segment since gene rich regions replicate early. Together, these results indicate that the PNC is associated with a chromatin locus or loci (Norton et al. 2008b).

In summary, PNCs are unique nuclear substructures that form in malignant cells. PNCs are highly enriched with newly synthesized pol III RNAs and are also concentrated with RNA binding proteins primarily indicated in pol II RNA metabolism. The PNC is nucleated upon a yet to be identified DNA locus or loci. These findings lead to a working model, in which the formation of the PNC is a cellular physiological adjustment to the malignant state during cancer development. The enrichment of pol III transcripts and RNA binding proteins suggests that these novel protein-RNA complexes in PNCs may play a role in the regulation of gene expression of pol III origins during transformation. Studies are underway to characterize the novel RNPs that are associated with the PNC and their role in pol III gene expression in cancer cells.

1.5 Evidence for the Coordination of Function Between Nuclear Subdomains in Tumour Suppression

Many of the nuclear bodies have physical associations, and frequently occur juxtaposed beside each other. The PNC, as its name would suggest, is found adjacent to the nucleolus (Huang et al. 1997). Cajal bodies are also frequently found at the nucleolar periphery, can move to and from the nucleolus (Platani et al. 2000), and have even been found within the nucleoli in human breast carcinoma cells (Ochs et al. 1994). Furthermore, gems, cleavage bodies, and PML NBs can colocalize with or lie adjacent to Cajal bodies (Dundr and Misteli 2001; Spector 2001). Various RNA and protein components of each of these bodies are also known to shuttle between structures. For example, snoRNAs are first transported to Cajal bodies for modification before they reach the nucleolus, and the Cajal body marker protein coilin can also be present in the nucleolus (Morris 2008), and the nucleolar protein fibrillarin can also shuttle to Cajal bodies (Dundr et al. 2004). Another example is the nucleolar protein nucleolin, which has also been found in the PNC (Pollock and Huang 2009). Finally, following nucleolar disassembly initiated by transcriptional inhibition, PML and Sp100 from PML NBs, and coilin from Cajal bodies form caps on the segregated nucleolar components (Shav-Tal et al. 2005). These data support a model in which the nuclear bodies function in an interrelated manner to accomplish their respective tasks. For example, snoRNA maturation in the Cajal bodies ultimately supports ribosome biogenesis in the nucleolus, and snRNA maturation and assembly into snRNPs in the nucleolus and Cajal bodies is required for the splicing function of the nuclear speckles (Dundr and Misteli 2001; Raska et al. 2006; Morris 2008).

This integration of nuclear body function also extends into their contributions to cancer biology. For example, the overlapping role of the nucleolus and Cajal bodies in ribonucleoprotein maturation also includes telomerase biogenesis and regulation. In cancer cells, telomerase is localized to Cajal bodies, with the telomerase reverse transcriptase (Tomlinson et al. 2008) and the novel telomerase holoenzyme subunit TCAB1 (telomerase Cajal body protein 1) (Venteicher et al. 2009) being required for the recruitment of the enzyme to this site. As telomerase maturation is also associated with the nucleolus, and active telomerase holoenzyme includes the nucleolar protein dyskerin, a functional telomerase enzyme may require modifications carried out at either or both the nucleolus and the Cajal body (Wong et al. 2002; Raska et al. 2006; Theimer et al. 2007; Derenzini et al. 2009). PML IV, one of the isoforms of PML, and a component of the PML NBs, is a negative regulator of telomerase through its interaction with TERT (Oh et al. 2009). Interestingly, in telomerase-negative tumours, PML NBs are associated with an alternative mechanism of telomere maintenance known as alternative lengthening of telomeres (ALT). In ALT cells, a subset of PML NBs associate with telomeric DNA, and these bodies are known as ALT-associated PML NBs (APBs) (reviewed in [Dellaire and Bazett-Jones 2004]). Therefore, at least three types of nuclear bodies can function to prevent telomere shortening in cancer cells, which allows these cells to divide indefinitely.

The nucleolus and PML NBs also exhibit substantial overlap in transient constituents involved in the regulation of genome stability and the DNA damage response. For example, the BLM protein, which is RecQ family helicase, is found in PML NBs except during S phase when it relocalizes to the nucleolus with another RecQ family helicase, WRN, potentially to help stabilize the replication of the rDNA repeats (Yankiwski et al. 2000; Schawalder et al. 2003). As discussed above, mutations in the *BLM* and *WRN* genes cause Bloom syndrome and Werner syndrome, respectively, which are human genomic instability syndromes that are also characterized by cancer predisposition (Hanada and Hickson 2007). Therefore, shuttling of the BLM protein between the PML NBs and the nucleolus may be important for regulating its function in genome maintenance. Other DNA damage response proteins associated with both the nucleolus and PML NBs include the critical replication-damage signaling kinase ATR (Andersen et al. 2002), the replication factor C-like clamp loader Rad17 (Chang et al. 1999); Rad50, a constituent of the damage sensing MRN complex (Andersen et al. 2002); and Rad52, which functions in homologous recombination repair (Liu et al. 1999). Rad17, Rad52, and ATR are associated with PML NBs in ALT cell lines (Barr et al. 2003; Dellaire and Bazett-Jones 2004), whereas Rad50 has been show to associate (as a part of the MRN complex) with PML NBs both in undamaged cells and at late repair foci (reviewed in [Dellaire and Bazett-Jones 2004]). Rad17 is normally localized to the nucleolus, but is released following UV-induced DNA damage (Chang et al. 1999), and Rad52 specifically localizes to nucleoli during S phase (Liu et al. 1999). Therefore, shuttling of these DNA damage proteins between nucleoli and PML NBs may represent: (1) multiple layers of regulation, including sequestering proteins until their functions are required; and/or (2) targeting of repair proteins to the two bodies for similar functions, such as aiding the replication of the rDNA repeats at the nucleoli and the telomere repeats at APBs.

Finally, the regulation of the p53 protein provides another important example of the functional overlap between nuclear bodies in cancer biology, as both PML NBs and the nucleolus have critical functions in the regulation of this fundamental tumour suppressor (Fig. 1.2). Depending on the activating signal (which can include oncogene expression or DNA damage), p53 can be stabilized at PML NBs through acetylation by the acetyltransferases CBP (Ferbeyre et al. 2000) or TIP60 (Wu et al. 2009), through phosphorylation by the kinases Chk2 (Louria-Hayon et al. 2003), CK1 (Alsheich-Bartok et al. 2008) or HIPK2 (D'Orazi et al. 2002), and potentially through de-ubiquitination by ubiquitin specific protease 7 (USP7) (Everett et al. 1997; Li et al. 2002). While the PML NBs provide a site for post-translational modification of p53, the nucleolus acts as an upstream stress sensor for p53. In times of nucleolar stress (when RNA polymerase I transcription is inhibited), the nucleolus disassembles, releasing nucleolar proteins into the nucleoplasm. A growing list of them, including ARF, NPM, nucleostemin, nucleolin, and the ribosomal proteins RPL5, RPL11, RPL23, and RPS7 help stabilize p53 through interactions with MDM2 and/or p53 (Montanaro et al. 2008; Zhang and Lu 2009). Furthermore, the nucleolus can also act as a site for MDM2 sequestration. Following DNA damage or oncogene expression, ARF can sequester MDM2 in the nucleolus, again preventing its interaction with p53 (Weber et al. 1999; Khan et al. 2004; Dias et al. 2006; Saporita et al. 2007). Finally, the functional link between the PML NBs

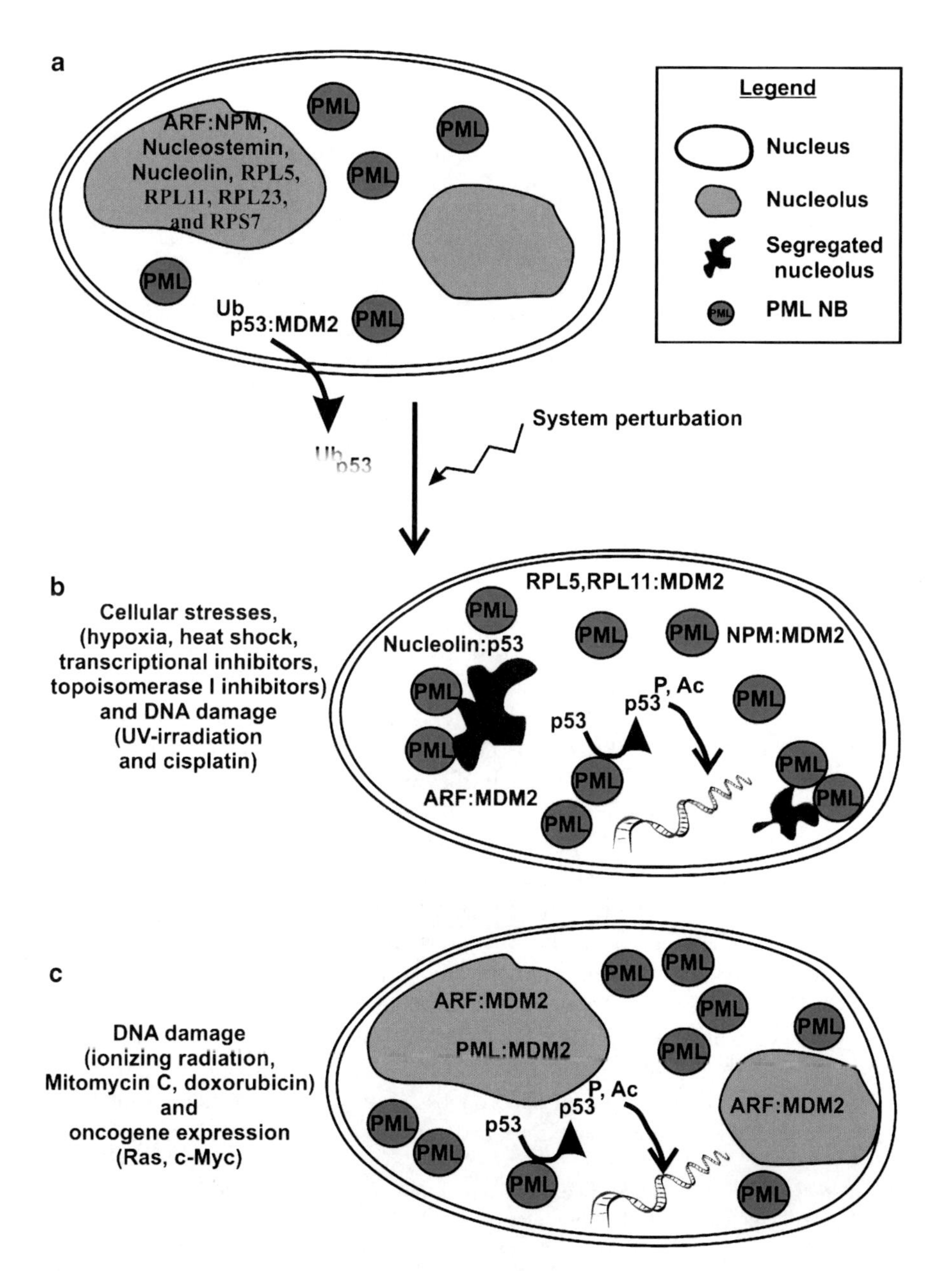

Fig. 1.2 Regulation of the p53 tumour suppressor by nuclear bodies. (**a**) In an unstressed cell, p53 levels are kept low by the action of MDM2. MDM2 is an E3 ligase that ubiquitinates (Ub) p53, targeting it for degradation. (**b**) Following cellular stress, including DNA damage that causes RNA polymerase I inhibition, the nucleolus disassembles. This disassembly releases ARF, NPM, nucleostemin, nucleolin, and the ribosomal proteins RPL5, RPL11, RPL23, and RPS7 into the nucleoplasm, where they help stabilize p53 through interactions with MDM2 and/or p53. p53 is also stabilized by phosphorylation (P) and acetylation (Ac) at PML NBs by the kinases Chk2, CK1 and HIPK2, and by the acetyltransferases CBP and TIP60. Stabilized and activated p53 can then upregulate its target genes, which include the G1/S regulator p21, and the pro-apoptotic genes Noxa, Puma, Bid, and Bax. (**c**) Following oncogene expression or certain forms of DNA damage (such as those caused by ionizing radiation, Mitomycin C, or doxorubicin), MDM2 is sequestered in the nucleolus by ARF and/or PML. p53 is then activated as above

and the nucleolus in p53 regulation is perhaps best exemplified by the observation that PML can relocalize to the nucleolus and also help sequester MDM2 there following doxorubicin treatment (Bernardi et al. 2004). Therefore, the nucleolus and the PML NBs can function independently and in concert to help ensure that p53 is stabilized and activated when it needs to be.

The nucleus is a dynamic, interconnected web of chromatin and nuclear bodies. DNA replication, cell cycle regulation, and DNA repair must be and are integrated with the ongoing processes of RNA transcription, RNA processing, ribonucleoprotein complex formation, and ribosome biogenesis. Therefore, it is not surprising that a single perturbation to the system can induce wide-spread consequences across the entire nuclear landscape. As such, no nuclear body is truly isolated from its neighbours.

1.6 Concluding Remarks

Our current understanding of nuclear body function would not have been possible without the advent of advanced microscopy techniques, including fluorescence microscopy, live-cell imaging, and electron spectroscopic imaging (Dellaire et al. 2004; Olson and Dundr 2005; Handwerger and Gall 2006; Hernandez-Verdun 2006b; Trinkle-Mulcahy and Lamond 2007). Interactions between proteins, nuclear body response to stimuli, and nuclear trafficking can now be visualized and analyzed in vivo, and the nuclear localization of almost any protein, RNA or gene can be mapped. Not surprisingly, these types of analyses led to the identification of many of the nuclear bodies, and helped establish new functions for well-known nuclear bodies such as the nucleolus. As such, the morphological analysis of cancer cells can be carried out at a level of detail that would not have been possible 20 years ago. Instead of merely associating certain cellular phenotypes with cancer cells, we now have the capabilities to begin to directly relate structural alterations with functional changes within the cell. As such, our knowledge of the roles that nuclear bodies play in tumour suppression and oncogenesis is increasing rapidly, providing new cancer biomarkers (for diagnosis, prognosis, remission status monitoring, treatment efficacy, and treatment selection), and new targets for cancer therapies.

Acknowledgements G.D. is the Cameron Scientist in Cancer Biology of the Dalhousie Cancer Research Program and a Canadian Institutes of Health Research (CIHR) New Investigator. This work is funded in part by operating grants from Nova Scotia Health Research Foundation (NSHRF) (2007–3348) and the CIHR (MOP-84260). K.L.C. is supported by postdoctoral fellowships from the NSHRF and the Killam Trusts. We would also like to apologize to those colleagues whom we could only cite indirectly in this review.

References

Alcalay M, Tomassoni L, Colombo E, Stoldt S, Grignani F, Fagioli M, Szekely L, Helin K, Pelicci PG (1998) The promyelocytic leukemia gene product (PML) forms stable complexes with the retinoblastoma protein. Mol Cell Biol 18:1084–1093

Alsheich-Bartok O, Haupt S, Alkalay-Snir I, Saito S, Appella E, Haupt Y (2008) PML enhances the regulation of p53 by CK1 in response to DNA damage. Oncogene 27:3653–3661
Andersen JS, Lyon CE, Fox AH, Leung AK, Lam YW, Steen H, Mann M, Lamond AI (2002) Directed proteomic analysis of the human nucleolus. Curr Biol 12:1–11
Arabi A, Wu S, Ridderstrale K, Bierhoff H, Shiue C, Fatyol K, Fahlen S, Hydbring P, Soderberg O, Grummt I, Larsson LG, Wright AP (2005) c-Myc associates with ribosomal DNA and activates RNA polymerase I transcription. Nat Cell Biol 7:303–310
Badhai J, Frojmark AS, Davey EJ, Schuster J, Dahl N (2009) Ribosomal protein S19 and S24 insufficiency cause distinct cell cycle defects in Diamond-blackfan anemia. Biochim Biophys Acta 1792: 1036–1042
Bailey D, O'Hare P (2002) Herpes simplex virus 1 ICP0 co-localizes with a SUMO-specific protease. J Gen Virol 83:2951–2964
Bailey SM, Murnane JP (2006) Telomeres, chromosome instability and cancer. Nucleic Acids Res 34:2408–2417
Balmain A, Gray J, Ponder B (2003) The genetics and genomics of cancer. Nat Genet 33(Suppl):238–244
Barr SM, Leung CG, Chang EE, Cimprich KA (2003) ATR kinase activity regulates the intranuclear translocation of ATR and RPA following ionizing radiation. Curr Biol 13:1047–1051
Bartkova J, Rezaei N, Liontos M, Karakaidos P, Kletsas D, Issaeva N, Vassiliou LV, Kolettas E, Niforou K, Zoumpourlis VC, Takaoka M, Nakagawa H, Tort F, Fugger K, Johansson F, Sehested M, Andersen CL, Dyrskjot L, Orntoft T, Lukas J, Kittas C, Helleday T, Halazonetis TD, Bartek J, Gorgoulis VG (2006) Oncogene-induced senescence is part of the tumorigenesis barrier imposed by DNA damage checkpoints. Nature 444:633–637
Bernardi R, Guernah I, Jin D, Grisendi S, Alimonti A, Teruya-Feldstein J, Cordon-Cardo C, Simon MC, Rafii S, Pandolfi PP (2006) PML inhibits HIF-1alpha translation and neoangiogenesis through repression of mTOR. Nature 442:779–785
Bernardi R, Pandolfi PP (2007) Structure, dynamics and functions of promyelocytic leukaemia nuclear bodies. Nat Rev Mol Cell Biol 8:1006–1016
Bernardi R, Papa A, Pandolfi PP (2008) Regulation of apoptosis by PML and the PML-NBs. Oncogene 27:6299–6312
Bernardi R, Scaglioni PP, Bergmann S, Horn HF, Vousden KH, Pandolfi PP (2004) PML regulates p53 stability by sequestering Mdm2 to the nucleolus. Nat Cell Biol 6:665–672
Bernassola F, Oberst A, Melino G, Pandolfi PP (2005) The promyelocytic leukaemia protein tumour suppressor functions as a transcriptional regulator of p63. Oncogene 24:6982–6986
Bernassola F, Salomoni P, Oberst A, Di Como CJ, Pagano M, Melino G, Pandolfi PP (2004) Ubiquitin-dependent degradation of p73 is inhibited by PML. J Exp Med 199:1545–1557
Bierhoff H, Dundr M, Michels AA, Grummt I (2008) Phosphorylation by casein kinase 2 facilitates rRNA gene transcription by promoting dissociation of TIF-IA from elongating RNA polymerase I. Mol Cell Biol 28:4988–4998
Bischof O, Kirsh O, Pearson M, Itahana K, Pelicci PG, Dejean A (2002) Deconstructing PML-induced premature senescence. Embo J 21:3358–3369
Bischof O, Nacerddine K, Dejean A (2005) Human papillomavirus oncoprotein E7 targets the promyelocytic leukemia protein and circumvents cellular senescence via the Rb and p53 tumor suppressor pathways. Mol Cell Biol 25:1013–1024
Boe SO, Haave M, Jul-Larsen A, Grudic A, Bjerkvig R, Lonning PE (2006) Promyelocytic leukemia nuclear bodies are predetermined processing sites for damaged DNA. J Cell Sci 119:3284–3295
Boisvert FM, Hendzel MJ, Masson JY, Richard S (2005) Methylation of MRE11 regulates its nuclear compartmentalization. Cell Cycle 4:981–989
Boisvert FM, van Koningsbruggen S, Navascues J, Lamond AI (2007) The multifunctional nucleolus. Nat Rev Mol Cell Biol 8:574–585
Bradshaw PS, Stavropoulos DJ, Meyn MS (2005) Human telomeric protein TRF2 associates with genomic double-strand breaks as an early response to DNA damage. Nat Genet 37:193–197
Budde A, Grummt I (1999) p53 represses ribosomal gene transcription. Oncogene 18:1119–1124

Bullwinkel J, Baron-Luhr B, Ludemann A, Wohlenberg C, Gerdes J, Scholzen T (2006) Ki-67 protein is associated with ribosomal RNA transcription in quiescent and proliferating cells. J Cell Physiol 206:624–635

Burikhanov R, Zhao Y, Goswami A, Qiu S, Schwarze SR, Rangnekar VM (2009) The tumor suppressor Par-4 activates an extrinsic pathway for apoptosis. Cell 138:377–388

Campisi J, D'Adda di Fagagna F (2007) Cellular senescence: when bad things happen to good cells. Nat Rev Mol Cell Biol 8:729–740

Carbone R, Pearson M, Minucci S, Pelicci PG (2002) PML NBs associate with the hMre11 complex and p53 at sites of irradiation induced DNA damage. Oncogene 21:1633–1640

Castle JC, Zhang C, Shah JK, Kulkarni AV, Kalsotra A, Cooper TA, Johnson JM (2008) Expression of 24, 426 human alternative splicing events and predicted cis regulation in 48 tissues and cell lines. Nat Genet 40:1416–1425

Chang F, Syrjanen S, Kurvinen K, Syrjanen K (1993) The p53 tumor suppressor gene as a common cellular target in human carcinogenesis. Am J Gastroenterol 88:174–186

Chang HY, Nishitoh H, Yang X, Ichijo H, Baltimore D (1998) Activation of apoptosis signal-regulating kinase 1 (ASK1) by the adapter protein Daxx. Science 281:1860–1863

Chang MS, Sasaki H, Campbell MS, Kraeft SK, Sutherland R, Yang CY, Liu Y, Auclair D, Hao L, Sonoda H, Ferland LH, Chen LB (1999) HRad17 colocalizes with NHP2L1 in the nucleolus and redistributes after UV irradiation. J Biol Chem 274:36544–36549

Charette SJ, Lavoie JN, Lambert H, Landry J (2000) Inhibition of Daxx-mediated apoptosis by heat shock protein 27. Mol Cell Biol 20:7602–7612

Chelbi-Alix MK, de The H (1999) Herpes virus induced proteasome-dependent degradation of the nuclear bodies-associated PML and Sp100 proteins. Oncogene 18:935–941

Cheng Z, Ke Y, Ding X, Wang F, Wang H, Wang W, Ahmed K, Liu Z, Xu Y, Aikhionbare F, Yan H, Liu J, Xue Y, Yu J, Powell M, Liang S, Wu Q, Reddy SE, Hu R, Huang H, Jin C, Yao X (2008) Functional characterization of TIP60 sumoylation in UV-irradiated DNA damage response. Oncogene 27:931–941

Ciarmatori S, Scott PH, Sutcliffe JE, McLees A, Alzuherri HM, Dannenberg JH, te Riele H, Grummt I, Voit R, White RJ (2001) Overlapping functions of the pRb family in the regulation of rRNA synthesis. Mol Cell Biol 21:5806–5814

Cohen SB, Graham ME, Lovrecz GO, Bache N, Robinson PJ, Reddel RR (2007) Protein composition of catalytically active human telomerase from immortal cells. Science 315:1850–1853

Collins K (2006) The biogenesis and regulation of telomerase holoenzymes. Nat Rev Mol Cell Biol 7:484–494

Colombo E, Marine JC, Danovi D, Falini B, Pelicci PG (2002) Nucleophosmin regulates the stability and transcriptional activity of p53. Nat Cell Biol 4:529–533

Condemine W, Takahashi Y, Zhu J, Puvion-Dutilleul F, Guegan S, Janin A, de The H (2006) Characterization of endogenous human promyelocytic leukemia isoforms. Cancer Res 66:6192–6198

Cortez D, Guntuku S, Qin J, Elledge SJ (2001) ATR and ATRIP: partners in checkpoint signaling. Science 294:1713–1716

D'Orazi G, Cecchinelli B, Bruno T, Manni I, Higashimoto Y, Saito S, Gostissa M, Coen S, Marchetti A, Del Sal G, Piaggio G, Fanciulli M, Appella E, Soddu S (2002) Homeodomain-interacting protein kinase-2 phosphorylates p53 at Ser 46 and mediates apoptosis. Nat Cell Biol 4:11–19

Dai MS, Lu H (2008) Crosstalk between c-Myc and ribosome in ribosomal biogenesis and cancer. J Cell Biochem 105:670–677

Dai MS, Sun XX, Lu H (2008) Aberrant expression of nucleostemin activates p53 and induces cell cycle arrest via inhibition of MDM2. Mol Cell Biol 28:4365–4376

Daniely Y, Borowiec JA (2000) Formation of a complex between nucleolin and replication protein A after cell stress prevents initiation of DNA replication. J Cell Biol 149:799–810

Daniely Y, Dimitrova DD, Borowiec JA (2002) Stress-dependent nucleolin mobilization mediated by p53-nucleolin complex formation. Mol Cell Biol 22:6014–6022

de Stanchina E, Querido E, Narita M, Davuluri RV, Pandolfi PP, Ferbeyre G, Lowe SW (2004) PML is a direct p53 target that modulates p53 effector functions. Mol Cell 13:523–535

de The H, Lavau C, Marchio A, Chomienne C, Degos L, Dejean A (1991) The PML-RAR alpha fusion mRNA generated by the t(15;17) translocation in acute promyelocytic leukemia encodes a functionally altered RAR. Cell 66:675–684
DeGregori J (2002) The genetics of the E2F family of transcription factors: shared functions and unique roles. Biochim Biophys Acta 1602:131–150
Delcuve GP, Rastegar M, Davie JR (2009) Epigenetic control. J Cell Physiol 219:243–250
Dellaire G, Bazett-Jones DP (2004) PML nuclear bodies: dynamic sensors of DNA damage and cellular stress. Bioessays 26:963–977
Dellaire G, Bazett-Jones DP (2007) Beyond repair foci: subnuclear domains and the cellular response to DNA damage. Cell Cycle 6:1864–1872
Dellaire G, Ching RW, Ahmed K, Jalali F, Tse KC, Bristow RG, Bazett-Jones DP (2006a) Promyelocytic leukemia nuclear bodies behave as DNA damage sensors whose response to DNA double-strand breaks is regulated by NBS1 and the kinases ATM, Chk2, and ATR. J Cell Biol 175:55–66
Dellaire G, Ching RW, Dehghani H, Ren Y, Bazett-Jones DP (2006b) The number of PML nuclear bodies increases in early S phase by a fission mechanism. J Cell Sci 119:1026–1033
Dellaire G, Eskiw CH, Dehghani H, Ching RW, Bazett-Jones DP (2006c) Mitotic accumulations of PML protein contribute to the re-establishment of PML nuclear bodies in G1. J Cell Sci 119:1034–1042
Dellaire G, Farrall R, Bickmore WA (2003) The Nuclear Protein Database (NPD): sub-nuclear localisation and functional annotation of the nuclear proteome. Nucleic Acids Res 31:328–330
Dellaire G, Kepkay R, Bazett-Jones DP (2009) High resolution imaging of changes in the structure and spatial organization of chromatin, gamma-H2A.X and the MRN complex within etoposide-induced DNA repair foci. Cell Cycle 8:3750–3769
Dellaire G, Nisman R, Bazett-Jones DP (2004) Correlative light and electron spectroscopic imaging of chromatin in situ. Methods Enzymol 375:456–478
Derenzini M (2000) The AgNORs. Micron 31:117–120
Derenzini M, Ceccarelli C, Santini D, Taffurelli M, Trere D (2004) The prognostic value of the AgNOR parameter in human breast cancer depends on the pRb and p53 status. J Clin Pathol 57:755–761
Derenzini M, Montanaro L, Trere D (2009) What the nucleolus says to a tumour pathologist. Histopathology 54:753–762
Dias SS, Milne DM, Meek DW (2006) c-Abl phosphorylates Hdm2 at tyrosine 276 in response to DNA damage and regulates interaction with ARF. Oncogene 25:6666–6671
Dowling RJ, Topisirovic I, Fonseca BD, Sonenberg N (2009) Dissecting the role of mTOR: lessons from mTOR inhibitors. Biochim Biophys Acta
Drygin D, Siddiqui-Jain A, O'Brien S, Schwaebe M, Lin A, Bliesath J, Ho CB, Proffitt C, Trent K, Whitten JP, Lim JK, Von Hoff D, Anderes K, Rice WG (2009) Anticancer activity of CX-3543: a direct inhibitor of rRNA biogenesis. Cancer Res 69:7653–7661
Dundr M, Hebert MD, Karpova TS, Stanek D, Xu H, Shpargel KB, Meier UT, Neugebauer KM, Matera AG, Misteli T (2004) In vivo kinetics of Cajal body components. J Cell Biol 164:831–842
Dundr M, Misteli T (2001) Functional architecture in the cell nucleus. Biochem J 356:297–310
Dyck JA, Maul GG, Miller WH Jr, Chen JD, Kakizuka A, Evans RM (1994) A novel macromolecular structure is a target of the promyelocyte-retinoic acid receptor oncoprotein. Cell 76:333–343
Dyck JA, Warrell RP Jr, Evans RM, Miller WH Jr (1995) Rapid diagnosis of acute promyelocytic leukemia by immunohistochemical localization of PML/RAR-alpha protein. Blood 86:862–867
Dyson N (1998) The regulation of E2F by pRB-family proteins. Genes Dev 12:2245–2262
Eladad S, Ye TZ, Hu P, Leversha M, Beresten S, Matunis MJ, Ellis NA (2005) Intra-nuclear trafficking of the BLM helicase to DNA damage-induced foci is regulated by SUMO modification. Hum Mol Genet 14:1351–1365
Elias JM (1997) Cell proliferation indexes: a biomarker in solid tumors. Biotech Histochem 72:78–85

Emmott E, Hiscox JA (2009) Nucleolar targeting: the hub of the matter. EMBO Rep 10:231–238
Everett RD (2001) DNA viruses and viral proteins that interact with PML nuclear bodies. Oncogene 20:7266–7273
Everett RD, Freemont P, Saitoh H, Dasso M, Orr A, Kathoria M, Parkinson J (1998) The disruption of ND10 during herpes simplex virus infection correlates with the Vmw110- and proteasome-dependent loss of several PML isoforms. J Virol 72:6581–6591
Everett RD, Meredith M, Orr A, Cross A, Kathoria M, Parkinson J (1997) A novel ubiquitin-specific protease is dynamically associated with the PML nuclear domain and binds to a herpesvirus regulatory protein. Embo J 16:1519–1530
Ferbeyre G, de Stanchina E, Querido E, Baptiste N, Prives C, Lowe SW (2000) PML is induced by oncogenic ras and promotes premature senescence. Genes Dev 14:2015–2027
Fogal V, Gostissa M, Sandy P, Zacchi P, Sternsdorf T, Jensen K, Pandolfi PP, Will H, Schneider C, Del Sal G (2000) Regulation of p53 activity in nuclear bodies by a specific PML isoform. Embo J 19:6185–6195
Fu C, Ahmed K, Ding H, Ding X, Lan J, Yang Z, Miao Y, Zhu Y, Shi Y, Zhu J, Huang H, Yao X (2005) Stabilization of PML nuclear localization by conjugation and oligomerization of SUMO-3. Oncogene 24:5401–5413
Gao C, Ho CC, Reineke E, Lam M, Cheng X, Stanya KJ, Liu Y, Chakraborty S, Shih HM, Kao HY (2008) Histone deacetylase 7 promotes PML sumoylation and is essential for PML nuclear body formation. Mol Cell Biol 28:5658–5667
Gao H, Chen XB, McGowan CH (2003) Mus81 endonuclease localizes to nucleoli and to regions of DNA damage in human S-phase cells. Mol Biol Cell 14:4826–4834
Genovese C, Trani D, Caputi M, Claudio PP (2006) Cell cycle control and beyond: emerging roles for the retinoblastoma gene family. Oncogene 25:5201–5209
German J, Ellis NA, Proytcheva M (1996) Bloom's syndrome. XIX. Cytogenetic and population evidence for genetic heterogeneity. Clin Genet 49:223–231
Ghetti A, Pinol-Roma S, Michael WM, Morandi C, Dreyfuss G (1992) hnRNP I, the polypyrimidine tract-binding protein: distinct nuclear localization and association with hnRNAs. Nucleic Acids Res 20:3671–3678
Gilbert N, Gilchrist S, Bickmore WA (2005) Chromatin organization in the mammalian nucleus. Int Rev Cytol 242:283–336
Gjerset RA (2006) DNA damage, p14ARF, nucleophosmin (NPM/B23), and cancer. J Mol Histol 37:239–251
Gong L, Millas S, Maul GG, Yeh ET (2000) Differential regulation of sentrinized proteins by a novel sentrin-specific protease. J Biol Chem 275:3355–3359
Gonzalez L, Freije JM, Cal S, Lopez-Otin C, Serrano M, Palmero I (2006) A functional link between the tumour suppressors ARF and p33ING1. Oncogene 25:5173–5179
Goodpasture C, Bloom SE (1975) Visualization of nucleolar organizer regions im mammalian chromosomes using silver staining. Chromosoma 53:37–50
Gostissa M, Hengstermann A, Fogal V, Sandy P, Schwarz SE, Scheffner M, Del Sal G (1999) Activation of p53 by conjugation to the ubiquitin-like protein SUMO-1. Embo J 18:6462–6471
Grandori C, Gomez-Roman N, Felton-Edkins ZA, Ngouenet C, Galloway DA, Eisenman RN, White RJ (2005) c-Myc binds to human ribosomal DNA and stimulates transcription of rRNA genes by RNA polymerase I. Nat Cell Biol 7:311–318
Gresko E, Ritterhoff S, Sevilla-Perez J, Roscic A, Frobius K, Kotevic I, Vichalkovski A, Hess D, Hemmings BA, Schmitz ML (2009) PML tumor suppressor is regulated by HIPK2-mediated phosphorylation in response to DNA damage. Oncogene 28:698–708
Guha M, Altieri DC (2009) Survivin as a global target of intrinsic tumor suppression networks. Cell Cycle 8:2708–2710
Guo A, Salomoni P, Luo J, Shih A, Zhong S, Gu W, Pandolfi PP (2000) The function of PML in p53-dependent apoptosis. Nat Cell Biol 2:730–736
Gurrieri C, Capodieci P, Bernardi R, Scaglioni PP, Nafa K, Rush LJ, Verbel DA, Cordon-Cardo C, Pandolfi PP (2004) Loss of the tumor suppressor PML in human cancers of multiple histologic origins. J Natl Cancer Inst 96:269–279

Hall MP, Huang S, Black DL (2004) Differentiation-induced colocalization of the KH-type splicing regulatory protein with polypyrimidine tract binding protein and the c-src pre-mRNA. Mol Biol Cell 15:774–786
Hanada K, Hickson ID (2007) Molecular genetics of RecQ helicase disorders. Cell Mol Life Sci 64:2306–2322
Handwerger KE, Gall JG (2006) Subnuclear organelles: new insights into form and function. Trends Cell Biol 16:19–26
Harbour JW, Dean DC (2000) The Rb/E2F pathway: expanding roles and emerging paradigms. Genes Dev 14:2393–2409
Harvey M, McArthur MJ, Montgomery CA Jr, Butel JS, Bradley A, Donehower LA (1993) Spontaneous and carcinogen-induced tumorigenesis in p53-deficient mice. Nat Genet 5:225–229
Hayakawa F, Abe A, Kitabayashi I, Pandolfi PP, Naoe T (2008) Acetylation of PML is involved in histone deacetylase inhibitor-mediated apoptosis. J Biol Chem 283:24420–24425
Hayakawa F, Privalsky ML (2004) Phosphorylation of PML by mitogen-activated protein kinases plays a key role in arsenic trioxide-mediated apoptosis. Cancer Cell 5:389–401
Heix J, Vente A, Voit R, Budde A, Michaelidis TM, Grummt I (1998) Mitotic silencing of human rRNA synthesis: inactivation of the promoter selectivity factor SL1 by cdc2/cyclin B-mediated phosphorylation. Embo J 17:7373–7381
Heliot L, Mongelard F, Klein C, O'Donohue MF, Chassery JM, Robert-Nicoud M, Usson Y (2000) Nonrandom distribution of metaphase AgNOR staining patterns on human acrocentric chromosomes. J Histochem Cytochem 48:13–20
Henras AK, Soudet J, Gerus M, Lebaron S, Caizergues-Ferrer M, Mougin A, Henry Y (2008) The post-transcriptional steps of eukaryotic ribosome biogenesis. Cell Mol Life Sci 65:2334–2359
Hernandez-Verdun D (2006a) The nucleolus: a model for the organization of nuclear functions. Histochem Cell Biol 126:135–148
Hernandez-Verdun D (2006b) Nucleolus: from structure to dynamics. Histochem Cell Biol 125:127–137
Hofmann TG, Moller A, Sirma H, Zentgraf H, Taya Y, Droge W, Will H, Schmitz ML (2002) Regulation of p53 activity by its interaction with homeodomain-interacting protein kinase-2. Nat Cell Biol 4:1–10
Hofmann TG, Stollberg N, Schmitz ML, Will H (2003) HIPK2 regulates transforming growth factor-beta-induced c-Jun NH(2)-terminal kinase activation and apoptosis in human hepatoma cells. Cancer Res 63:8271–8277
Huang S (2000) Review: perinucleolar structures. J Struct Biol 129:233–240
Huang S, Deerinck TJ, Ellisman MH, Spector DL (1997) The dynamic organization of the perinucleolar compartment in the cell nucleus. J Cell Biol 137:965–974
Huang S, Deerinck TJ, Ellisman MH, Spector DL (1998) The perinucleolar compartment and transcription. J Cell Biol 143:35–47
Hussain S, Benavente SB, Nascimento E, Dragoni I, Kurowski A, Gillich A, Humphreys P, Frye M (2009) The nucleolar RNA methyltransferase Misu (NSun2) is required for mitotic spindle stability. J Cell Biol 186:27–40
Huttelmaier S, Illenberger S, Grosheva I, Rudiger M, Singer RH, Jockusch BM (2001) Raver1, a dual compartment protein, is a ligand for PTB/hnRNPI and microfilament attachment proteins. J Cell Biol 155:775–786
Imai Y, Kimura T, Murakami A, Yajima N, Sakamaki K, Yonehara S (1999) The CED-4-homologous protein FLASH is involved in Fas-mediated activation of caspase-8 during apoptosis. Nature 398:777–785
Ishov AM, Sotnikov AG, Negorev D, Vladimirova OV, Neff N, Kamitani T, Yeh ET, Strauss JF, Maul GG (1999) PML is critical for ND10 formation and recruits the PML-interacting protein daxx to this nuclear structure when modified by SUMO-1. J Cell Biol 147:221–234
Iwakuma T, Lozano G (2003) MDM2, an introduction. Mol Cancer Res 1:993–1000
James MJ, Zomerdijk JC (2004) Phosphatidylinositol 3-kinase and mTOR signaling pathways regulate RNA polymerase I transcription in response to IGF-1 and nutrients. J Biol Chem 279:8911–8918

Jensen K, Shiels C, Freemont PS (2001) PML protein isoforms and the RBCC/TRIM motif. Oncogene 20:7223–7233
Johnson FB, Marciniak RA, Guarente L (1998) Telomeres, the nucleolus and aging. Curr Opin Cell Biol 10:332–338
Junttila MR, Evan GI (2009) p53–a Jack of all trades but master of none. Nat Rev Cancer 9:821–829
Kakizuka A, Miller WH Jr, Umesono K, Warrell RP Jr, Frankel SR, Murty VV, Dmitrovsky E, Evans RM (1991) Chromosomal translocation t(15;17) in human acute promyelocytic leukemia fuses RAR alpha with a novel putative transcription factor, PML. Cell 66:663–674
Kaldis P, Aleem E (2005) Cell cycle sibling rivalry: Cdc2 vs. Cdk2. Cell Cycle 4:1491–1494
Kamath RV, Thor AD, Wang C, Edgerton SM, Slusarczyk A, Leary DJ, Wang J, Wiley EL, Jovanovic B, Wu Q, Nayar R, Kovarik P, Shi F, Huang S (2005) Perinucleolar compartment prevalence has an independent prognostic value for breast cancer. Cancer Res 65:246–253
Karmakar P, Bohr VA (2005) Cellular dynamics and modulation of WRN protein is DNA damage specific. Mech Ageing Dev 126:1146–1158
Kawai T, Akira S, Reed JC (2003) ZIP kinase triggers apoptosis from nuclear PML oncogenic domains. Mol Cell Biol 23:6174–6186
Kemp CJ, Wheldon T, Balmain A (1994) p53-deficient mice are extremely susceptible to radiation-induced tumorigenesis. Nat Genet 8:66–69
Khan S, Guevara C, Fujii G, Parry D (2004) p14ARF is a component of the p53 response following ionizing irradiation of normal human fibroblasts. Oncogene 23:6040–6046
Killian A, Le Meur N, Sesboue R, Bourguignon J, Bougeard G, Gautherot J, Bastard C, Frebourg T, Flaman JM (2004) Inactivation of the RRB1-Pescadillo pathway involved in ribosome biogenesis induces chromosomal instability. Oncogene 23:8597–8602
Kinzler KW, Vogelstein B (1997) Cancer-susceptibility genes. Gatekeepers and caretakers. Nature 386(761):763
Kirwan M, Dokal I (2009) Dyskeratosis congenita, stem cells and telomeres. Biochim Biophys Acta 1792:371–379
Kitagawa D, Kajiho H, Negishi T, Ura S, Watanabe T, Wada T, Ichijo H, Katada T, Nishina H (2006) Release of RASSF1C from the nucleus by Daxx degradation links DNA damage and SAPK/JNK activation. Embo J 25:3286–3297
Koken MH, Puvion-Dutilleul F, Guillemin MC, Viron A, Linares-Cruz G, Stuurman N, de Jong L, Szostecki C, Calvo F, Chomienne C et al (1994) The t(15;17) translocation alters a nuclear body in a retinoic acid-reversible fashion. Embo J 13:1073–1083
Krieghoff-Henning E, Hofmann TG (2008) Role of nuclear bodies in apoptosis signalling. Biochim Biophys Acta 1783:2185–2194
Kruse JP, Gu W (2009) Modes of p53 regulation. Cell 137:609–622
Kumagai A, Lee J, Yoo HY, Dunphy WG (2006) TopBP1 activates the ATR-ATRIP complex. Cell 124:943–955
Kurki S, Latonen L, Laiho M (2003) Cellular stress and DNA damage invoke temporally distinct Mdm2, p53 and PML complexes and damage-specific nuclear relocalization. J Cell Sci 116:3917–3925
Kurki S, Peltonen K, Latonen L, Kiviharju TM, Ojala PM, Meek D, Laiho M (2004) Nucleolar protein NPM interacts with HDM2 and protects tumor suppressor protein p53 from HDM2-mediated degradation. Cancer Cell 5:465–475
Lai HK, Borden KL (2000) The promyelocytic leukemia (PML) protein suppresses cyclin D1 protein production by altering the nuclear cytoplasmic distribution of cyclin D1 mRNA. Oncogene 19:1623–1634
Lallemand-Breitenbach V, Jeanne M, Benhenda S, Nasr R, Lei M, Peres L, Zhou J, Zhu J, Raught B, de The H (2008) Arsenic degrades PML or PML-RARalpha through a SUMO-triggered RNF4/ubiquitin-mediated pathway. Nat Cell Biol 10:547–555
Lallemand-Breitenbach V, Zhu J, Puvion F, Koken M, Honore N, Doubeikovsky A, Duprez E, Pandolfi PP, Puvion E, Freemont P, de The H (2001) Role of promyelocytic leukemia (PML)

sumolation in nuclear body formation, 11S proteasome recruitment, and As2O3-induced PML or PML/retinoic acid receptor alpha degradation. J Exp Med 193:1361–1371
Lane AA, Chabner BA (2009) Histone deacetylase inhibitors in cancer therapy. J Clin Oncol 27:5459–5468
Langley E, Pearson M, Faretta M, Bauer UM, Frye RA, Minucci S, Pelicci PG, Kouzarides T (2002) Human SIR2 deacetylates p53 and antagonizes PML/p53-induced cellular senescence. Embo J 21:2383–2396
Lapi E, Di Agostino S, Donzelli S, Gal H, Domany E, Rechavi G, Pandolfi PP, Givol D, Strano S, Lu X, Blandino G (2008) PML, YAP, and p73 are components of a proapoptotic autoregulatory feedback loop. Mol Cell 32:803–814
Laronne A, Rotkopf S, Hellman A, Gruenbaum Y, Porter AC, Brandeis M (2003) Synchronization of interphase events depends neither on mitosis nor on cdk1. Mol Biol Cell 14:3730–3740
Le XF, Vallian S, Mu ZM, Hung MC, Chang KS (1998) Recombinant PML adenovirus suppresses growth and tumorigenicity of human breast cancer cells by inducing G1 cell cycle arrest and apoptosis. Oncogene 16:1839–1849
Lee B, Matera AG, Ward DC, Craft J (1996) Association of RNase mitochondrial RNA processing enzyme with ribonuclease P in higher ordered structures in the nucleolus: a possible coordinate role in ribosome biogenesis. P Nat Acad Sci U S A 93:11471–11476
Levine AJ (1997) p53, the cellular gatekeeper for growth and division. Cell 88:323–331
Levine AJ, Oren M (2009) The first 30 years of p53: growing ever more complex. Nat Rev Cancer 9:749–758
Li H, Leo C, Zhu J, Wu X, O'Neil J, Park EJ, Chen JD (2000) Sequestration and inhibition of Daxx-mediated transcriptional repression by PML. Mol Cell Biol 20:1784–1796
Li L, He D, He H, Wang X, Zhang L, Luo Y, Nan X (2006) Overexpression of PML induced apoptosis in bladder cancer cell by caspase dependent pathway. Cancer Lett 236:259–268
Li M, Brooks CL, Kon N, Gu W (2004) A dynamic role of HAUSP in the p53-Mdm2 pathway. Mol Cell 13:879–886
Li M, Chen D, Shiloh A, Luo J, Nikolaev AY, Qin J, Gu W (2002) Deubiquitination of p53 by HAUSP is an important pathway for p53 stabilization. Nature 416:648–653
Li SJ, Hochstrasser M (1999) A new protease required for cell-cycle progression in yeast. Nature 398:246–251
Li Z, Boone D, Hann SR (2008) Nucleophosmin interacts directly with c-Myc and controls c-Myc-induced hyperproliferation and transformation. P Nat Acad Sci U S A 105:18794–18799
Lima CD, Reverter D (2008) Structure of the human SENP7 catalytic domain and poly-SUMO deconjugation activities for SENP6 and SENP7. J Biol Chem 283:32045–32055
Lin CY, Navarro S, Reddy S, Comai L (2006a) CK2-mediated stimulation of Pol I transcription by stabilization of UBF-SL1 interaction. Nucleic Acids Res 34:4752–4766
Lin DY, Huang YS, Jeng JC, Kuo HY, Chang CC, Chao TT, Ho CC, Chen YC, Lin TP, Fang HI, Hung CC, Suen CS, Hwang MJ, Chang KS, Maul GG, Shih HM (2006b) Role of SUMO-interacting motif in Daxx SUMO modification, subnuclear localization, and repression of sumoylated transcription factors. Mol Cell 24:341–354
Lin HK, Bergmann S, Pandolfi PP (2004) Cytoplasmic PML function in TGF-beta signalling. Nature 431:205–211
Liu Y, Li M, Lee EY, Maizels N (1999) Localization and dynamic relocalization of mammalian Rad52 during the cell cycle and in response to DNA damage. Curr Biol 9:975–978
Lombard DB, Guarente L (2000) Nijmegen breakage syndrome disease protein and MRE11 at PML nuclear bodies and meiotic telomeres. Cancer Res 60:2331–2334
Louria-Hayon I, Alsheich-Bartok O, Levav-Cohen Y, Silberman I, Berger M, Grossman T, Matentzoglu K, Jiang YH, Muller S, Scheffner M, Haupt S, Haupt Y (2009) E6AP promotes the degradation of the PML tumor suppressor. Cell Death Differ 16:1156–1166
Louria-Hayon I, Grossman T, Sionov RV, Alsheich O, Pandolfi PP, Haupt Y (2003) The promyelocytic leukemia protein protects p53 from Mdm2-mediated inhibition and degradation. J Biol Chem 278:33134–33141

Louvet E, Junera HR, Berthuy I, Hernandez-Verdun D (2006) Compartmentation of the nucleolar processing proteins in the granular component is a CK2-driven process. Mol Biol Cell 17:2537–2546

Luo J, Su F, Chen D, Shiloh A, Gu W (2000) Deacetylation of p53 modulates its effect on cell growth and apoptosis. Nature 408:377–381

Ma H, Pederson T (2008) Nucleostemin: a multiplex regulator of cell-cycle progression. Trends Cell Biol 18:575–579

Maggi LB Jr, Weber JD (2005) Nucleolar adaptation in human cancer. Cancer Invest 23:599–608

Maiguel DA, Jones L, Chakravarty D, Yang C, Carrier F (2004) Nucleophosmin sets a threshold for p53 response to UV radiation. Mol Cell Biol 24:3703–3711

Mallette FA, Ferbeyre G (2007) The DNA damage signaling pathway connects oncogenic stress to cellular senescence. Cell Cycle 6:1831–1836

Mallette FA, Goumard S, Gaumont-Leclerc MF, Moiseeva O, Ferbeyre G (2004) Human fibroblasts require the Rb family of tumor suppressors, but not p53, for PML-induced senescence. Oncogene 23:91–99

Marciniak RA, Lombard DB, Johnson FB, Guarente L (1998) Nucleolar localization of the Werner syndrome protein in human cells. Proc Natl Acad Sci U S A 95:6887–6892

Marcos-Villar L, Lopitz-Otsoa F, Gallego P, Munoz-Fontela C, Gonzalez-Santamaria J, Campagna M, Shou-Jiang G, Rodriguez MS, Rivas C (2009) Kaposi's sarcoma-associated herpesvirus protein LANA2 disrupts PML oncogenic domains and inhibits PML-mediated transcriptional repression of the survivin gene. J Virol 83:8849–8858

Mastrocola AS, Heinen CD (2009) Nuclear reorganization of DNA mismatch repair proteins in response to DNA damage. DNA Repair (Amst)

Matera AG, Frey MR, Margelot K, Wolin SL (1995) A perinucleolar compartment contains several RNA polymerase III transcripts as well as the polypyrimidine tract-binding protein, hnRNP I. J Cell Biol 129:1181–1193

Matera AG, Izaguire-Sierra M, Praveen K, Rajendra TK (2009) Nuclear bodies: random aggregates of sticky proteins or crucibles of macromolecular assembly? Dev Cell 17:639–647

Mayer C, Bierhoff H, Grummt I (2005) The nucleolus as a stress sensor: JNK2 inactivates the transcription factor TIF-IA and down-regulates rRNA synthesis. Genes Dev 19:933–941

Mayer C, Grummt I (2005) Cellular stress and nucleolar function. Cell Cycle 4:1036–1038

Mayer C, Zhao J, Yuan X, Grummt I (2004) mTOR-dependent activation of the transcription factor TIF-IA links rRNA synthesis to nutrient availability. Genes Dev 18:423–434

Meinecke I, Cinski A, Baier A, Peters MA, Dankbar B, Wille A, Drynda A, Mendoza H, Gay RE, Hay RT, Ink B, Gay S, Pap T (2007) Modification of nuclear PML protein by SUMO-1 regulates Fas-induced apoptosis in rheumatoid arthritis synovial fibroblasts. Proc Natl Acad Sci U S A 104:5073–5078

Menendez D, Inga A, Resnick MA (2009) The expanding universe of p53 targets. Nat Rev Cancer 9:724–737

Meng L, Lin T, Tsai RY (2008) Nucleoplasmic mobilization of nucleostemin stabilizes MDM2 and promotes G2-M progression and cell survival. J Cell Sci 121:4037–4046

Milovic-Holm K, Krieghoff E, Jensen K, Will H, Hofmann TG (2007) FLASH links the CD95 signaling pathway to the cell nucleus and nuclear bodies. Embo J 26:391–401

Mirzoeva OK, Petrini JH (2001) DNA damage-dependent nuclear dynamics of the Mre11 complex. Mol Cell Biol 21:281–288

Moll UM, Petrenko O (2003) The MDM2-p53 interaction. Mol Cancer Res 1:1001–1008

Mongelard F, Bouvet P (2007) Nucleolin: a multiFACeTed protein. Trends Cell Biol 17:80–86

Montanaro L, Trere D, Derenzini M (2008) Nucleolus, ribosomes, and cancer. Am J Pathol 173:301–310

Morris GE (2008) The Cajal body. Biochim Biophys Acta 1783:2108–2115

Mu ZM, Le XF, Vallian S, Glassman AB, Chang KS (1997) Stable overexpression of PML alters regulation of cell cycle progression in HeLa cells. Carcinogenesis 18:2063–2069

Mukhopadhyay D, Ayaydin F, Kolli N, Tan SH, Anan T, Kametaka A, Azuma Y, Wilkinson KD, Dasso M (2006) SUSP1 antagonizes formation of highly SUMO2/3-conjugated species. J Cell Biol 174:939–949

Mukhopadhyay D, Dasso M (2007) Modification in reverse: the SUMO proteases. Trends Biochem Sci 32:286–295
Naka K, Ikeda K, Motoyama N (2002) Recruitment of NBS1 into PML oncogenic domains via interaction with SP100 protein. Biochem Biophys Res Commun 299:863–871
Naoe T, Suzuki T, Kiyoi H, Urano T (2006) Nucleophosmin: a versatile molecule associated with hematological malignancies. Cancer Sci 97:963–969
Narita M, Nunez S, Heard E, Narita M, Lin AW, Hearn SA, Spector DL, Hannon GJ, Lowe SW (2003) Rb-mediated heterochromatin formation and silencing of E2F target genes during cellular senescence. Cell 113:703–716
Norton JT, Pollock CB, Wang C, Schink JC, Kim JJ, Huang S (2008a) Perinucleolar compartment prevalence is a phenotypic pancancer marker of malignancy. Cancer 113:861–869
Norton JT, Wang C, Gjidoda A, Henry RW, Huang S (2008b) The perinucleolar compartment is directly associated with DNA. J Biol Chem
Ochs RL, Stein TW Jr, Tan EM (1994) Coiled bodies in the nucleolus of breast cancer cells. J Cell Sci 107(Pt 2):385–399
Oh W, Ghim J, Lee EW, Yang MR, Kim ET, Ahn JH, Song J (2009) PML-IV functions as a negative regulator of telomerase by interacting with TERT. J Cell Sci 122:2613–2622
Olson MO (2004) Sensing cellular stress: another new function for the nucleolus? Sci STKE 2004:pe10
Olson MO, Dundr M (2005) The moving parts of the nucleolus. Histochem Cell Biol 123:203–216
Opresko PL, von Kobbe C, Laine JP, Harrigan J, Hickson ID, Bohr VA (2002) Telomere-binding protein TRF2 binds to and stimulates the Werner and Bloom syndrome helicases. J Biol Chem 277:41110–41119
Oza P, Peterson CL (2010) Opening the DNA repair toolbox: localization of DNA double strand breaks to the nuclear periphery. Cell Cycle 9:43–49
Pan D, Zhu Q, Luo K (2009) SnoN functions as a tumour suppressor by inducing premature senescence. Embo J 28:3500–3513
Pearson M, Carbone R, Sebastiani C, Cioce M, Fagioli M, Saito S, Higashimoto Y, Appella E, Minucci S, Pandolfi PP, Pelicci PG (2000) PML regulates p53 acetylation and premature senescence induced by oncogenic Ras. Nature 406:207–210
Pederson T (1998) The plurifunctional nucleolus. Nucleic Acids Res 26:3871–3876
Petrie K, Zelent A (2008) Marked for death. Nat Cell Biol 10:507–509
Pich A, Chiusa L, Audisio E, Marmont F (1998) Nucleolar organizer region counts predict complete remission, remission duration, and survival in adult acute myelogenous leukemia patients. J Clin Oncol 16:1512–1518
Pich A, Chiusa L, Margaria E (2000) Prognostic relevance of AgNORs in tumor pathology. Micron 31:133–141
Platani M, Goldberg I, Swedlow JR, Lamond AI (2000) In vivo analysis of Cajal body movement, separation, and joining in live human cells. J Cell Biol 151:1561–1574
Politz JC, Polena I, Trask I, Bazett-Jones DP, Pederson T (2005) A nonribosomal landscape in the nucleolus revealed by the stem cell protein nucleostemin. Mol Biol Cell 16:3401–3410
Pollock C, Huang S (2009) The perinucleolar compartment. J Cell Biochem 107:189–193
Pombo A, Cuello P, Schul W, Yoon JB, Roeder RG, Cook PR, Murphy S (1998) Regional and temporal specialization in the nucleus: a transcriptionally-active nuclear domain rich in PTF, Oct1 and PIKA antigens associates with specific chromosomes early in the cell cycle. Embo J 17:1768–1778
Raska I, Shaw PJ, Cmarko D (2006) New insights into nucleolar architecture and activity. Int Rev Cytol 255:177–235
Reineke EL, Lam M, Liu Q, Liu Y, Stanya KJ, Chang KS, Means AR, Kao HY (2008) Degradation of the tumor suppressor PML by Pin1 contributes to the cancer phenotype of breast cancer MDA-MB-231 cells. Mol Cell Biol 28:997–1006
Rideau AP, Gooding C, Simpson PJ, Monie TP, Lorenz M, Huttelmaier S, Singer RH, Matthews S, Curry S, Smith CWJ (2006) A peptide motif in Raver1 mediates splicing repression by interaction with the PTB RRM2 domain. Nat Struct Mol Biol 13:839–848

Rippe K (2007) Dynamic organization of the cell nucleus. Curr Opin Genet Dev 17:373–380
Rodier F, Campisi J, Bhaumik D (2007) Two faces of p53: aging and tumor suppression. Nucleic Acids Res 35:7475–7484
Romanova L, Grand A, Zhang L, Rayner S, Katoku-Kikyo N, Kellner S, Kikyo N (2009a) Critical role of nucleostemin in pre-rRNA processing. J Biol Chem 284:4968–4977
Romanova L, Kellner S, Katoku-Kikyo N and Kikyo N (2009b) Novel role of nucleostemin in the maintenance of nucleolar architecture and integrity of small nucleolar ribonucleoproteins and the telomerase complex. J Biol Chem
Roussel P, Hernandez-Verdun D (1994) Identification of Ag-NOR proteins, markers of proliferation related to ribosomal gene activity. Exp Cell Res 214:465–472
Roussigne M, Cayrol C, Clouaire T, Amalric F, Girard JP (2003) THAP1 is a nuclear proapoptotic factor that links prostate-apoptosis-response-4 (Par-4) to PML nuclear bodies. Oncogene 22:2432–2442
Rubbi CP, Milner J (2003) Disruption of the nucleolus mediates stabilization of p53 in response to DNA damage and other stresses. Embo J 22:6068–6077
Ruggero D, Grisendi S, Piazza F, Rego E, Mari F, Rao PH, Cordon-Cardo C, Pandolfi PP (2003) Dyskeratosis congenita and cancer in mice deficient in ribosomal RNA modification. Science 299:259–262
Ruggero D, Pandolfi PP (2003) Does the ribosome translate cancer? Nat Rev Cancer 3:179–192
Sachdev S, Bruhn L, Sieber H, Pichler A, Melchior F, Grosschedl R (2001) PIASy, a nuclear matrix-associated SUMO E3 ligase, represses LEF1 activity by sequestration into nuclear bodies. Genes Dev 15:3088–3103
Saijo Y, Sato G, Usui K, Sato M, Sagawa M, Kondo T, Minami Y, Nukiwa T (2001) Expression of nucleolar protein p120 predicts poor prognosis in patients with stage I lung adenocarcinoma. Ann Oncol 12:1121–1125
Saitoh N, Uchimura Y, Tachibana T, Sugahara S, Saitoh H, Nakao M (2006) In situ SUMOylation analysis reveals a modulatory role of RanBP2 in the nuclear rim and PML bodies. Exp Cell Res 312:1418–1430
Salomoni P (2009) Stemming out of a new PML era? Cell Death Differ 16:1083–1092
Salomoni P, Bernardi R, Bergmann S, Changou A, Tuttle S, Pandolfi PP (2005) The promyelocytic leukemia protein PML regulates c-Jun function in response to DNA damage. Blood 105:3686–3690
Salomoni P, Ferguson BJ, Wyllie AH, Rich T (2008) New insights into the role of PML in tumour suppression. Cell Res 18:622–640
Salomoni P, Khelifi AF (2006) Daxx: death or survival protein? Trends Cell Biol 16:97–104
Salomoni P, Pandolfi PP (2002) The role of PML in tumor suppression. Cell 108:165–170
Sanz MA, Grimwade D, Tallman MS, Lowenberg B, Fenaux P, Estey EH, Naoe T, Lengfelder E, Buchner T, Dohner H, Burnett A K, Lo-Coco F (2008) Guidelines on the management of acute promyelocytic leukemia: recommendations from an expert panel on behalf of the European LeukemiaNet. Blood
Sanz MM, Proytcheva M, Ellis NA, Holloman WK, German J (2000) BLM, the Bloom's syndrome protein, varies during the cell cycle in its amount, distribution, and co-localization with other nuclear proteins. Cytogenet Cell Genet 91:217–223
Saporita AJ, Maggi LB Jr, Apicelli AJ, Weber JD (2007) Therapeutic targets in the ARF tumor suppressor pathway. Curr Med Chem 14:1815–1827
Sawicka K, Bushell M, Spriggs KA, Willis AE (2008) Polypyrimidine-tract-binding protein: a multifunctional RNA-binding protein. Biochem Soc Trans 036:641–647
Saxena A, Rorie CJ, Dimitrova D, Daniely Y, Borowiec JA (2006) Nucleolin inhibits Hdm2 by multiple pathways leading to p53 stabilization. Oncogene 25:7274–7288
Scaglioni PP, Yung TM, Cai LF, Erdjument-Bromage H, Kaufman AJ, Singh B, Teruya-Feldstein J, Tempst P, Pandolfi PP (2006) A CK2-dependent mechanism for degradation of the PML tumor suppressor. Cell 126:269–283
Schawalder J, Paric E, Neff NF (2003) Telomere and ribosomal DNA repeats are chromosomal targets of the bloom syndrome DNA helicase. BMC Cell Biol 4:15

Scheffner M, Huibregtse JM, Vierstra RD, Howley PM (1993) The HPV-16 E6 and E6-AP complex functions as a ubiquitin-protein ligase in the ubiquitination of p53. Cell 75:495–505

Scheper GC, Parra JL, Wilson M, Van Kollenburg B, Vertegaal AC, Han ZG, Proud CG (2003) The N and C termini of the splice variants of the human mitogen-activated protein kinase-interacting kinase Mnk2 determine activity and localization. Mol Cell Biol 23:5692–5705

Scherl A, Coute Y, Deon C, Calle A, Kindbeiter K, Sanchez JC, Greco A, Hochstrasser D, Diaz JJ (2002) Functional proteomic analysis of human nucleolus. Mol Biol Cell 13:4100–4109

Schmidt EV (1999) The role of c-myc in cellular growth control. Oncogene 18:2988–2996

Schober H, Ferreira H, Kalck V, Gehlen LR, Gasser SM (2009) Yeast telomerase and the SUN domain protein Mps3 anchor telomeres and repress subtelomeric recombination. Genes Dev 23:928–938

Scholzen T, Gerdes J (2000) The Ki-67 protein: from the known and the unknown. J Cell Physiol 182:311–322

Scott M, Boisvert FM, Vieyra D, Johnston RN, Bazett-Jones DP, Riabowol K (2001) UV induces nucleolar translocation of ING1 through two distinct nucleolar targeting sequences. Nucleic Acids Res 29:2052–2058

Shav-Tal Y, Blechman J, Darzacq X, Montagna C, Dye BT, Patton JG, Singer RH, Zipori D (2005) Dynamic sorting of nuclear components into distinct nucleolar caps during transcriptional inhibition. Mol Biol Cell 16:2395–2413

Shechter D, Costanzo V, Gautier J (2004) Regulation of DNA replication by ATR: signaling in response to DNA intermediates. DNA Repair (Amst) 3:901–908

Shen HM, Tergaonkar V (2009) NFkappaB signaling in carcinogenesis and as a potential molecular target for cancer therapy. Apoptosis 14:348–363

Shen TH, Lin HK, Scaglioni PP, Yung TM, Pandolfi PP (2006) The mechanisms of PML-nuclear body formation. Mol Cell 24:331–339

Shtutman M, Zhurinsky J, Oren M, Levina E, Ben-Ze'ev A (2002) PML is a target gene of beta-catenin and plakoglobin, and coactivates beta-catenin-mediated transcription. Cancer Res 62:5947–5954

Sirri V, Roussel P, Hernandez-Verdun D (1999) The mitotically phosphorylated form of the transcription termination factor TTF-1 is associated with the repressed rDNA transcription machinery. J Cell Sci 112(Pt 19):3259–3268

Sirri V, Urcuqui-Inchima S, Roussel P, Hernandez-Verdun D (2008) Nucleolus: the fascinating nuclear body. Histochem Cell Biol 129:13–31

Soignet SL, Maslak P, Wang ZG, Jhanwar S, Calleja E, Dardashti LJ, Corso D, DeBlasio A, Gabrilove J, Scheinberg DA, Pandolfi PP, Warrell RP Jr (1998) Complete remission after treatment of acute promyelocytic leukemia with arsenic trioxide. N Engl J Med 339:1341–1348

Song MS, Salmena L, Carracedo A, Egia A, Lo-Coco F, Teruya-Feldstein J, Pandolfi PP (2008) The deubiquitinylation and localization of PTEN are regulated by a HAUSP-PML network. Nature 455:813–817

Soussi T, Lozano G (2005) p53 mutation heterogeneity in cancer. Biochem Biophys Res Commun 331:834–842

Spector DL (2001) Nuclear domains. J Cell Sci 114:2891–2893

Stagno D'Alcontres M, Mendez-Bermudez A, Foxon JL, Royle NJ, Salomoni P (2007) Lack of TRF2 in ALT cells causes PML-dependent p53 activation and loss of telomeric DNA. J Cell Biol 179:855–867

Stankovic-Valentin N, Deltour S, Seeler J, Pinte S, Vergoten G, Guerardel C, Dejean A, Leprince D (2007) An acetylation/deacetylation-SUMOylation switch through a phylogenetically conserved psiKXEP motif in the tumor suppressor HIC1 regulates transcriptional repression activity. Mol Cell Biol 27:2661–2675

Stefanovsky V, Langlois F, Gagnon-Kugler T, Rothblum LI, Moss T (2006) Growth factor signaling regulates elongation of RNA polymerase I transcription in mammals via UBF phosphorylation and r-chromatin remodeling. Mol Cell 21:629–639

Sternsdorf T, Guldner HH, Szostecki C, Grotzinger T, Will H (1995) Two nuclear dot-associated proteins, PML and Sp100, are often co-autoimmunogenic in patients with primary biliary cirrhosis. Scand J Immunol 42:257–268

Sternsdorf T, Jensen K, Will H (1997) Evidence for covalent modification of the nuclear dot-associated proteins PML and Sp100 by PIC1/SUMO-1. J Cell Biol 139:1621–1634
Storck S, Shukla M, Dimitrov S, Bouvet P (2007) Functions of the histone chaperone nucleolin in diseases. Subcell Biochem 41:125–144
Strano S, Monti O, Pediconi N, Baccarini A, Fontemaggi G, Lapi E, Mantovani F, Damalas A, Citro G, Sacchi A, Del Sal G, Levrero M, Blandino G (2005) The transcriptional coactivator Yes-associated protein drives p73 gene-target specificity in response to DNA Damage. Mol Cell 18:447–459
Strezoska Z, Pestov DG, Lau LF (2000) Bop1 is a mouse WD40 repeat nucleolar protein involved in 28S and 5. 8S RRNA processing and 60S ribosome biogenesis. Mol Cell Biol 20:5516–5528
Sugimoto M, Kuo ML, Roussel MF, Sherr CJ (2003) Nucleolar Arf tumor suppressor inhibits ribosomal RNA processing. Mol Cell 11:415–424
Sun H, Tu X, Prisco M, Wu A, Casiburi I, Baserga R (2003) Insulin-like growth factor I receptor signaling and nuclear translocation of insulin receptor substrates 1 and 2. Mol Endocrinol 17:472–486
Sun J, Xu H, Subramony SH, Hebert MD (2005) Interactions between coilin and PIASy partially link Cajal bodies to PML bodies. J Cell Sci 118:4995–5003
Sykes SM, Mellert HS, Holbert MA, Li K, Marmorstein R, Lane WS, McMahon SB (2006) Acetylation of the p53 DNA-binding domain regulates apoptosis induction. Mol Cell 24:841–851
Szostecki C, Guldner HH, Netter HJ, Will H (1990) Isolation and characterization of cDNA encoding a human nuclear antigen predominantly recognized by autoantibodies from patients with primary biliary cirrhosis. J Immunol 145:4338–4347
Takagi M, Absalon MJ, McLure KG, Kastan MB (2005) Regulation of p53 translation and induction after DNA damage by ribosomal protein L26 and nucleolin. Cell 123:49–63
Tang J, Qu LK, Zhang J, Wang W, Michaelson JS, Degenhardt YY, El-Deiry WS, Yang X (2006) Critical role for Daxx in regulating Mdm2. Nat Cell Biol 8:855–862
Tatham MH, Geoffroy MC, Shen L, Plechanovova A, Hattersley N, Jaffray EG, Palvimo JJ, Hay RT (2008) RNF4 is a poly-SUMO-specific E3 ubiquitin ligase required for arsenic-induced PML degradation. Nat Cell Biol 10:538–546
Theimer CA, Jady BE, Chim N, Richard P, Breece KE, Kiss T, Feigon J (2007) Structural and functional characterization of human telomerase RNA processing and cajal body localization signals. Mol Cell 27:869–881
Thomas G (2000) An encore for ribosome biogenesis in the control of cell proliferation. Nat Cell Biol 2:E71–E72
Tian X, Chen B, Liu X 2009. Telomere and telomerase as targets for Cancer therapy. Appl Biochem Biotechnol
Tomlinson RL, Abreu EB, Ziegler T, Ly H, Counter CM, Terns RM, Terns MP (2008) Telomerase reverse transcriptase is required for the localization of telomerase RNA to cajal bodies and telomeres in human cancer cells. Mol Biol Cell 19:3793–3800
Trabucchi M, Briata P, Garcia-Mayoral M, Haase AD, Filipowicz W, Ramos A, Gherzi R, Rosenfeld MG (2009) The RNA-binding protein KSRP promotes the biogenesis of a subset of microRNAs. Nature 459:1010–1014
Tran H, Brunet A, Griffith EC, Greenberg ME (2003) The many forks in FOXO's road. Sci STKE 2003:RE5
Trembley JH, Wang G, Unger G, Slaton J, Ahmed K (2009) Protein kinase CK2 in health and disease: CK2: a key player in cancer biology. Cell Mol Life Sci 66:1858–1867
Trere D (2000) AgNOR staining and quantification. Micron 31:127–131
Trere D, Ceccarelli C, Montanaro L, Tosti E, Derenzini M (2004) Nucleolar size and activity are related to pRb and p53 status in human breast cancer. J Histochem Cytochem 52:1601–1607
Trinkle-Mulcahy L, Lamond AI (2007) Toward a high-resolution view of nuclear dynamics. Science 318:1402–1407
Trotman LC, Alimonti A, Scaglioni PP, Koutcher JA, Cordon-Cardo C, Pandolfi PP (2006) Identification of a tumour suppressor network opposing nuclear Akt function. Nature 441:523–527

Tu X, Batta P, Innocent N, Prisco M, Casaburi I, Belletti B, Baserga R (2002) Nuclear translocation of insulin receptor substrate-1 by oncogenes and Igf-I. Effect on ribosomal RNA synthesis. J Biol Chem 277:44357–44365

Van Holde KE, Allen JR, Tatchell K, Weischet WO, Lohr D (1980) DNA-histone interactions in nucleosomes. Biophys J 32:271–282

Venteicher AS, Abreu EB, Meng Z, McCann KE, Terns RM, Veenstra TD, Terns MP, Artandi SE (2009) A human telomerase holoenzyme protein required for Cajal body localization and telomere synthesis. Science 323:644–648

Voit R, Grummt I (2001) Phosphorylation of UBF at serine 388 is required for interaction with RNA polymerase I and activation of rDNA transcription. Proc Natl Acad Sci U S A 98:13631–13636

Voit R, Hoffmann M, Grummt I (1999) Phosphorylation by G1-specific cdk-cyclin complexes activates the nucleolar transcription factor UBF. Embo J 18:1891–1899

Voit R, Schafer K, Grummt I (1997) Mechanism of repression of RNA polymerase I transcription by the retinoblastoma protein. Mol Cell Biol 17:4230–4237

Voit R, Schnapp A, Kuhn A, Rosenbauer H, Hirschmann P, Stunnenberg HG, Grummt I (1992) The nucleolar transcription factor mUBF is phosphorylated by casein kinase II in the C-terminal hyperacidic tail which is essential for transactivation. Embo J 11:2211–2218

Volarevic S, Stewart MJ, Ledermann B, Zilberman F, Terracciano L, Montini E, Grompe M, Kozma SC, Thomas G (2000) Proliferation, but not growth, blocked by conditional deletion of 40S ribosomal protein S6. Science 288:2045–2047

Wang C, Ivanov A, Chen L, Fredericks WJ, Seto E, Rauscher FJ, Chen J (2005) MDM2 interaction with nuclear corepressor KAP1 contributes to p53 inactivation. Embo J 24:3279–3290

Wang C, Politz JC, Pederson T, Huang S (2003) RNA polymerase III transcripts and the PTB protein are essential for the integrity of the perinucleolar compartment. Mol Biol Cell 14:2425–2435

Wang ZG, Delva L, Gaboli M, Rivi R, Giorgio M, Cordon-Cardo C, Grosveld F, Pandolfi PP (1998a) Role of PML in cell growth and the retinoic acid pathway. Science 279:1547–1551

Wang ZG, Ruggero D, Ronchetti S, Zhong S, Gaboli M, Rivi R, Pandolfi PP (1998b) PML is essential for multiple apoptotic pathways. Nat Genet 20:266–272

Wang ZY, Chen Z (2008) Acute promyelocytic leukemia: from highly fatal to highly curable. Blood 111:2505–2515

Weber JD, Taylor LJ, Roussel MF, Sherr CJ, Bar-Sagi D (1999) Nucleolar Arf sequesters Mdm2 and activates p53. Nat Cell Biol 1:20–26

Wong JM, Kusdra L, Collins K (2002) Subnuclear shuttling of human telomerase induced by transformation and DNA damage. Nat Cell Biol 4:731–736

Woo LL, Futami K, Shimamoto A, Furuichi Y, Frank KM (2006) The Rothmund-Thomson gene product RECQL4 localizes to the nucleolus in response to oxidative stress. Exp Cell Res 312:3443–3457

Wu A, Tu X, Prisco M, Baserga R (2005) Regulation of upstream binding factor 1 activity by insulin-like growth factor I receptor signaling. J Biol Chem 280:2863–2872

Wu G, Lee WH, Chen PL (2000) NBS1 and TRF1 colocalize at promyelocytic leukemia bodies during late S/G2 phases in immortalized telomerase-negative cells. Implication of NBS1 in alternative lengthening of telomeres. J Biol Chem 275:30618–30622

Wu Q, Hu H, Lan J, Emenari C, Wang Z, Chang KS, Huang H, Yao X (2009) PML3 Orchestrates the nuclear dynamics and function of TIP60. J Biol Chem 284:8747–8759

Wu WS, Xu ZX, Hittelman WN, Salomoni P, Pandolfi PP, Chang KS (2003) Promyelocytic leukemia protein sensitizes tumor necrosis factor alpha-induced apoptosis by inhibiting the NF-kappaB survival pathway. J Biol Chem 278:12294–12304

Xiao S, Scott F, Fierke CA, Engelke DR (2002) Eukaryotic ribonuclease P: a plurality of ribonucleoprotein enzymes. Ann Rev Biochem 71:165–189

Xu ZX, Timanova-Atanasova A, Zhao RX, Chang KS (2003) PML colocalizes with and stabilizes the DNA damage response protein TopBP1. Mol Cell Biol 23:4247–4256

Xu ZX, Zhao RX, Ding T, Tran TT, Zhang W, Pandolfi PP, Chang KS (2004) Promyelocytic leukemia protein 4 induces apoptosis by inhibition of survivin expression. J Biol Chem 279:1838–1844

Yamauchi M, Oka Y, Yamamoto M, Niimura K, Uchida M, Kodama S, Watanabe M, Sekine I, Yamashita S, Suzuki K (2008) Growth of persistent foci of DNA damage checkpoint factors is essential for amplification of G1 checkpoint signaling. DNA Repair (Amst) 7:405–417

Yang S, Jeong JH, Brown AL, Lee CH, Pandolfi PP, Chung JH, Kim MK (2006) Promyelocytic leukemia activates Chk2 by mediating Chk2 autophosphorylation. J Biol Chem 281:26645–26654

Yang S, Kuo C, Bisi JE, Kim MK (2002) PML-dependent apoptosis after DNA damage is regulated by the checkpoint kinase hCds1/Chk2. Nat Cell Biol 4:865–870

Yang X, Khosravi-Far R, Chang HY, Baltimore D (1997) Daxx, a novel Fas-binding protein that activates JNK and apoptosis. Cell 89:1067–1076

Yankiwski V, Marciniak RA, Guarente L, Neff NF (2000) Nuclear structure in normal and Bloom syndrome cells. Proc Natl Acad Sci U S A 97:5214–5219

Yates KE, Korbel GA, Shtutman M, Roninson IB, DiMaio D (2008) Repression of the SUMO-specific protease Senp1 induces p53-dependent premature senescence in normal human fibroblasts. Aging Cell 7:609–621

Yeager TR, Neumann AA, Englezou A, Huschtscha LI, Noble JR, Reddel RR (1999) Telomerase-negative immortalized human cells contain a novel type of promyelocytic leukemia (PML) body. Cancer Res 59:4175–4179

Yoon A, Peng G, Brandenburger Y, Zollo O, Xu W, Rego E, Ruggero D (2006) Impaired control of IRES-mediated translation in X-linked dyskeratosis congenita. Science 312:902–906

Yuan M, Tomlinson V, Lara R, Holliday D, Chelala C, Harada T, Gangeswaran R, Manson-Bishop C, Smith P, Danovi SA, Pardo O, Crook T, Mein CA, Lemoine NR, Jones LJ, Basu S (2008) Yes-associated protein (YAP) functions as a tumor suppressor in breast. Cell Death Differ 15:1752–1759

Zhai W, Comai L (2000) Repression of RNA polymerase I transcription by the tumor suppressor p53. Mol Cell Biol 20:5930–5938

Zhang Q, Hu CM, Yuan YS, He CH, Zhao Q, Liu NZ (2008) Expression of Mina53 and its significance in gastric carcinoma. Int J Biol Markers 23:83–88

Zhang R, Chen W, Adams PD (2007) Molecular dissection of formation of senescence-associated heterochromatin foci. Mol Cell Biol 27:2343–2358

Zhang R, Poustovoitov MV, Ye X, Santos HA, Chen W, Daganzo SM, Erzberger JP, Serebriiskii IG, Canutescu AA, Dunbrack RL, Pehrson JR, Berger JM, Kaufman PD, Adams PD (2005) Formation of MacroH2A-containing senescence-associated heterochromatin foci and senescence driven by ASF1a and HIRA. Dev Cell 8:19–30

Zhang S, Hemmerich P, Grosse F (2004) Nucleolar localization of the human telomeric repeat binding factor 2 (TRF2). J Cell Sci 117:3935–3945

Zhang Y, Lu H (2009) Signaling to p53: ribosomal proteins find their way. Cancer Cell 16:369–377

Zhao J, Yuan X, Frodin M, Grummt I (2003) ERK-dependent phosphorylation of the transcription initiation factor TIF-IA is required for RNA polymerase I transcription and cell growth. Mol Cell 11:405–413

Zhong S, Hu P, Ye TZ, Stan R, Ellis NA, Pandolfi PP (1999) A role for PML and the nuclear body in genomic stability. Oncogene 18:7941–7947

Zhong S, Muller S, Ronchetti S, Freemont PS, Dejean A, Pandolfi PP (2000a) Role of SUMO-1-modified PML in nuclear body formation. Blood 95:2748–2752

Zhong S, Salomoni P, Ronchetti S, Guo A, Ruggero D, Pandolfi PP (2000b) Promyelocytic leukemia protein (PML) and Daxx participate in a novel nuclear pathway for apoptosis. J Exp Med 191:631–640

Zhu J, Lallemand-Breitenbach V, de The H (2001) Pathways of retinoic acid- or arsenic trioxide-induced PML/RARalpha catabolism, role of oncogene degradation in disease remission. Oncogene 20:7257–7265

Zhu Z, Luo Z, Li Y, Ni C, Li H, Zhu M (2009) Human inhibitor of growth 1 inhibits hepatoma cell growth and influences p53 stability in a variant-dependent manner. Hepatology 49:504–512

Zou L, Elledge SJ (2003) Sensing DNA damage through ATRIP recognition of RPA-ssDNA complexes. Science 300:1542–1548

Chapter 2
Spatial Point Process Analysis of Promyelocytic Leukemia Nuclear Bodies

Philip P. Umande and David A. Stephens

Abstract There has been widespread interest in the nuclear body (NB) Promyelocytic leukemia (PML) because of its link to several human disorders, including Promyelocytic leukemia and AIDS. The notion of PML NB interaction with its surrounding and other NBs such as RNA Polymerase II (RNA Pol II) is of great importance as it can improve our understanding of the function of PML. In this paper, spatial point process methods are used to conduct multivariate analysis to assess the relationship between the spatial locations of PML NBs relative to RNA Pol II. We also propose a model for PML NB locations. By fitting a model to the PML NBs we are able to gain insight into how PML NBs are distributed across the nucleus in relation to themselves and the nuclear boundary.

Keywords Spatial Point Pattern • PML • RNA Pol II • Marked Point Process • Inhomogeneous Poisson Process • K-function

2.1 Introduction

The notion of Promyelocytic leukemia (PML) nuclear body (NB) interaction with its surrounding is one of great importance. Lanctot et al. (2007) have reported that gene expression is mediated by interaction between chromatin and protein complexes. Dellaire and Bazett-Jones (2004) have proposed that PML NBs are dynamic sensors of cellular stress, that associates with regions of DNA damage. Borden

P.P. Umande (✉)
Department of Mathematics, Imperial College, 180 Queens Gate, London SW7 2AZ, UK
e-mail: philip.umande00@imperial.ac.uk

D.A. Stephens
Department of Mathematics and Statistics, McGill University, 805 Sherbrooke Street West, Montreal QC, H3A 2K6, Canada

N.M. Adams and P.S. Freemont (eds.), *Advances in Nuclear Architecture*,
DOI 10.1007/978-90-481-9899-3_2,

(2002) has suggested that PML NBs tend to be near certain nuclear compartments such as Cajal/coiled bodies, cleavage bodies, and splicing speckles .

Knowledge of the PML NBs in relation to other structures may provide clues to PML NB functions. Furthermore, its relative location may give insight into what specific targets it regulates (Borden 2002). Following this, most reported strategies for assessing PML NB functions in essence are designed to answer questions relating to which nuclear structures the bodies are near to, what other macromolecules localize with the body, and the effects of disrupting the locations of the body (Borden 2002). Wang et al. (2004) analysed the correlation between the minimum locus-PML distances against their transcriptional activity to show that PML associate with transcriptionally active genomic regions.

The Imperial College Centre for Structural Biology has been able to provide images obtained via confocal microscopy. Such images provide the spatial locations (three-dimensional coordinate space) of PML and centroids of other nuclear bodies (RNA Polymerase II etc.) (see Fig. 2.1), and also the nuclear boundary (see Fig. 2.2). These data will be used for the quantitative analysis presented in this paper. Note that we are considering replicated data, that is, data for several cell nuclei representing multiple independently and identically distributed realisations of some spatial stochastic process. Replication can add complications to statistical inference (see Section 2.4 on modelling PML NB data), but is important as it provides further credibility to the outcome of the statistical analysis.

We shall carry out multivariate (or equivalently, *marked*) point pattern analysis so that we can provide some statistical evidence for biological ideas regarding

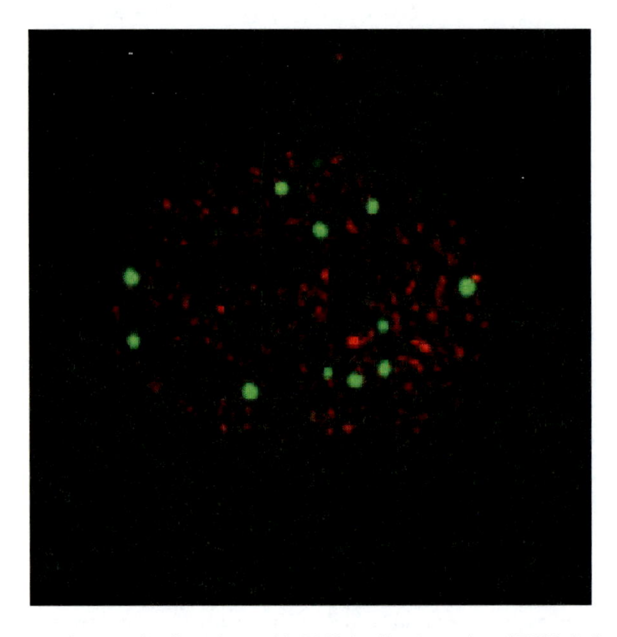

Fig. 2.1 Microscopy image of cell nucleus with PML bodies (*green*) and RNA-Polymerase II (*red*)

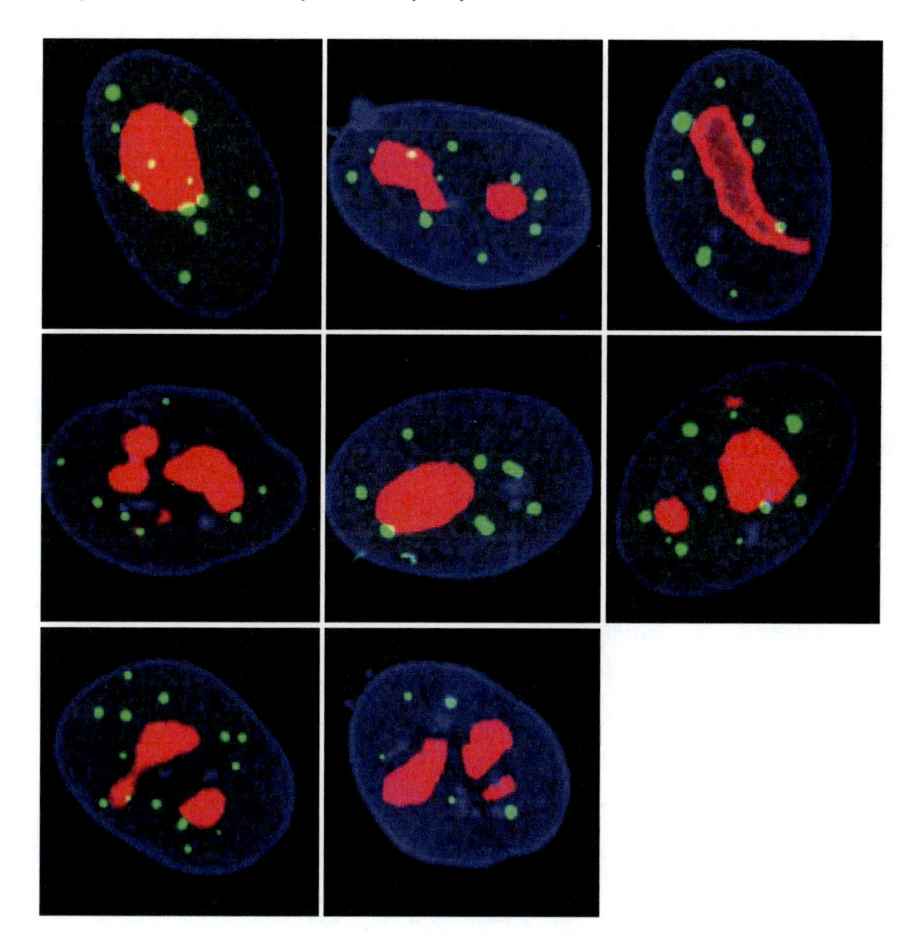

Fig. 2.2 PPSD2 data: Microscopy image of cell nucleus with PML bodies (*green*) and nucleoli (*red*), with nuclear lamina (*blue*)

PML NBs association with other nuclear bodies. Specifically, Borden (2002) stated that general transcription factors such as RNA Polymerase II do not colocalize with PML bodies. This is consistent with experimental results obtained by Xie and Pombo (2006) which supported the view that although PML NBs are present in transcriptionally active areas, they are not generally sites of polymerase II assembly. We shall attempt to use spatial point pattern analysis to confirm or otherwise these findings. We also explore and discuss how one might go about modelling the relationship between PML NB size with its positioning in the nuclear interior.

This paper is structured as follows. In the next section we will provide a background to point process theory and its application for multiple event types. We shall then go onto discuss how multivariate spatial analysis can be used for assessing the relationship between the spatial location of PML NBs relative to RNA Polymerase II. In the final section we discuss a possible approach for modelling PML NB locations before ending with concluding remarks.

2.2 Spatial Point Processes: Theory, Models and Statistics

The earliest discussions on spatial point processes date back to the early 1950s when used by Skellam (1952) and Thompson (1955) in statistical ecology. The work of Matérn in 1960, later re-published in 1986, has been viewed as pioneering (Cox and Isham 1980; Kemp 1988). More recently, Møller and Waagepetersen (2004) have provided more detailed accounts of modern spatial point process theory, statistics and models; see also Stoyan (2006).

We begin by introducing notation, definitions and some important concepts. These initial definitions and concepts are as those given by Cressie (1991), Karr (1991), Stoyan and Stoyan (1994), Stoyan et al. (1995), and Stoyan (2006). We begin by considering a collection of points observed within the nucleus, resulting in a data set having two or three coordinates if we utilize microscopic slices or the entire confocal image respectively. Throughout we denote locations by $\boldsymbol{x}=(x,y)$ or $\boldsymbol{x}=(x,y,z)$ and use subscripts $i=1,\ldots,n$ to denote the locations of n observations in the data set. We shall denote the 2-dimensional disk (3-dimensional ball) of radius r centered at $\boldsymbol{x}$ by $\mathcal{B}^d(\boldsymbol{x},r)$, although the superscript d will sometimes be omitted.

2.3 Spatial Point Process: Definitions and Calculations

We assume that the points – PMLs, genomic loci, other nuclear bodies stained in the experiment – observed in the images follow a *spatial point process*, that is, the points occur according to a random mechanism that we can characterize in terms of the distribution of the numbers of points that occur in disjoint spatial regions. For a spatial point process, denoted X, we write

$$X(B) = \sum_{x_i \in X} \mathbb{I}_B(\boldsymbol{x}_i),$$

where $\mathbb{I}_B(.)$ denotes the indicator function for set B, to indicate the count of the number of points of X observed in the region B. For set S, the notation $X(S)=n$ means that S contains n points of X. A *spatial point pattern* – a realisation of a spatial point process – is defined through the locations of points (Cressie 1991). We will at times make the assumption that the spatial point process X is *stationary* or *isotropic*; X is *stationary* (or equivalently, *homogeneous*) if it has the property that $X_{x'} = \{\boldsymbol{x}' + \boldsymbol{x} : \boldsymbol{x} \in X\}$ has the same distribution for all $\boldsymbol{x}' \in R^d$. Also, the point process X is said to be *isotropic* if it is invariant under rotation. Stationarity and isotropy can be very important assumptions in spatial point process analysis. Practitioners will at times (often implicitly) assume that stationarity holds (without carrying out formal tests for stationarity), or be content that it holds approximately so that certain point process techniques can be readily adopted (see Baddeley et al. 1992; Glasbey and Roberts 1997 as examples).

The *intensity measure*, Λ, of X is a point process characteristic analogous to the mean of real-valued random variables, that is defined as

$$\Lambda(B) = \mathbb{E}[X(B)] = \int \boldsymbol{x}(B)P(d\boldsymbol{x}) = \mathbb{E}\left[\sum_{x \in X} \mathbb{I}_B(\boldsymbol{x})\right]$$

that is, the expected number of points lying in the region B. In the *homogeneous* case it suffices to consider an *intensity*, λ since then $\Lambda(B) = \lambda \nu(B)$, where $\nu(B)$ is the area (or volume) of B. In general, we define the *kth moment measure* $\mu^{(k)}$ by

$$\mu^{(k)}(B_1 \times \cdots \times B_k) = \mathbb{E}\left[\prod_{i=1}^{k} X(B_i)\right]$$

for sets $B_1, \ldots, B_k$, and $\times$ denoting the Cartesian product. Stoyan and Stoyan (1994) provide a more detailed account on higher order moments, including geometrical interpretations.

2.4 Binomial and Poisson Point Processes

2.4.1 The Binomial Point Process

The points $\boldsymbol{x}_1 \ldots \boldsymbol{x}_n$ form a *Binomial point process*, $X_{\mathrm{Bin}(W,n)}$ in the set W if they are independently and uniformly distributed inside W, with

$$\mathbb{P}(\boldsymbol{x}_1 \in B_1, \ldots, \boldsymbol{x}_n \in B_n) = \mathbb{P}(\boldsymbol{x} \in B_1) \cdot \cdots \cdot \mathbb{P}(\boldsymbol{x} \in B_n) = \frac{\nu(B_1) \cdot \cdots \cdot \nu(B_n)}{\nu(W)^n}$$

for $B_1, \ldots, B_n$ subsets of W. The intensity, λ, of this process is given by

$$\lambda \;=\; \frac{np(B)}{\nu(B)} \quad \text{where} \quad p(A) = \frac{\nu(A)}{\nu(W)}, \quad A \subseteq W.$$

The simulation of virtually all spatial point processes in W requires simulating a binomial point process (n points uniformly inside W). Simulating n events uniformly inside W, a unit square or cube is straightforward; for a unit cube, one simply superimposes n independent uniform random points, $\boldsymbol{u}_1, \ldots, \boldsymbol{u}_n$ where $\boldsymbol{u}_i = (u_{i1}, u_{i2}, u_{i3})$, and $u_{ij} \sim \mathit{Uniform}(0, 1)$. Once simulated inside the unit cube, we can then apply scaling and translation in order to obtain a simulation inside any fixed cuboid. Coordinate transformation can be used for simulating uniformly inside a sphere. That is, if $\boldsymbol{u}=(u_1,u_2,u_3)$ is a uniform point in the unit cube, then $\boldsymbol{x}=(x_1,x_2,x_3)$ where

$$x_1 = R\sin(\theta)\cos(\phi) \qquad x_2 = R\sin(\theta)\sin(\phi) \qquad x_3 = R\cos(\theta)$$

and $R=u_1^{1/3}$, $\theta=\arccos(1-2u_2)$, and $\varphi=2\pi u_3$ is uniform inside the unit sphere. Once again, the relevant scaling can be applied for the case where simulation inside an ellipsoid is required. For less straightforward sets and irregularly shaped regions, W_0, rejection sampling (see for example Ripley 1987) can be used. This involves, for example, simulating uniformly inside $W \supset W_0$ and retaining the points that lie in W_0. Simulation is repeated until the desired number of points are obtained.

Data sets PPDS2 and PPDS3 provide the locations of points that constitute empty space inside the nucleus. We exploit this information as a means of simulating u points uniformly inside the nucleus. Specifically, for the u points $\boldsymbol{x}_1,\ldots,\boldsymbol{x}_u$ that are classified as empty space, n such points are chosen at random (without replacement). Random selection is made by equipping each of the $\boldsymbol{x}_1,\ldots,\boldsymbol{x}_u$ with a unique integer $k \in \{1,\ldots,u\}$. The point $\boldsymbol{x}$ is selected if the randomly chosen $y \in \{1,2,\ldots,u\}$ is its assigned integer. Occasionally, we attempt to simulate uniformly inside the nuclear interior by adopting methods for simulating uniformly inside a convex hull (Fishman 1996).

2.4.2 *The Homogeneous Poisson Point Process*

A default standard model for point patterns is the *homogeneous Poisson process*. The homogeneous (stationary) Poisson process X_p, is defined by the following postulates:

(i) For some constant $\lambda>0$, and set B, $X_p(B)$ follows a Poisson distribution with mean $\lambda\nu(B)$. The parameter λ is the intensity. For the three-dimensional case, this can be interpreted as the number of events per unit volume.

(ii) Given $X_p(B)=n$, the n events in B form an independent sample from the uniform distribution on B.

(iii) If $B_1,\ldots,B_s$ are disjoint sets then $X_p(B_1),\ldots,X_p(B_s)$ are independent Poisson random variables with mean $\lambda\nu(B_1),\ldots,\lambda\nu(B_s)$. Therefore

$$\mathbb{P}(X_p(B_1)=n_1,\ldots,X_p(B_s)=n_s) = \frac{\nu(B_1)^{n_1}\cdot\cdots\cdot\nu(B_s)^{n_s}}{n_1!\cdot\cdots\cdot n_s!}\lambda^{\sum_{i=1}^{s} n_i}\exp\left(-\sum_{i=1}^{s}\lambda\nu(B_i)\right)$$

These postulates are simultaneously the definition of *complete spatial randomness* (CSR). The *void* probabilities of X_p are given by $\mathbb{P}(X(B)=0)=\exp(-\lambda\nu(B))$. The first order moment Λ follows from property (i) and is given by

$$\Lambda(B) = \mathbb{E}[X(B)] = \lambda\nu(B)$$

The homogeneous Poisson spatial point process can be simulated inside W directly from the first two postulates. First we simulate $n \sim \text{Poisson}(\lambda \nu(B))$, then simulate n points uniformly inside W (that is, simulate $\tilde{x} \sim X_{\text{Bin}(W,n)}$).

2.4.3 Inhomogeneous Poisson Point Process

One of the most simple alternatives to the homogeneous Poisson point process is the *inhomogeneous Poisson point process*. The *inhomogeneous Poisson point process* is obtained by replacing the intensity λ by a spatially varying density $\lambda(\boldsymbol{x})$. Let Λ be a diffuse Radon measure on $\mathbb{R}^d$. An *inhomogeneous Poisson point process* is a point process possessing the following two properties:

(i) The number of events in a bounded B has a Poisson distribution with mean $\Lambda(B)$

$$\mathbb{P}(X(B)=n)=\frac{(\Lambda(B))^n}{n!}\exp\left(-\Lambda(B)\right), \quad \text{for} \quad n \in \{0,1,2,\ldots\}$$

(ii) The number of points in k disjoint sets form k independent random variables.

The function $\Lambda(B)$ can be written as

$$\Lambda(B)=\int_B \lambda(\boldsymbol{x})d\boldsymbol{x}$$

The function $\lambda(.)$ is the *intensity function* of the inhomogeneous point process. An inhomogeneous Poisson spatial point process in a set W can be simulated by using the rejection or *random thinning* algorithm of Lewis and Shedler (1976). The algorithm for an inhomogeneous Poisson process with intensity function $\lambda(\boldsymbol{x})$ is as follows: Simulate a homogeneous spatial Poisson point process of intensity λ_{max}, where λ_{max} is the largest intensity value over W. Then independently delete each point $\boldsymbol{x}_i$ with probability

$$1-\frac{\lambda(x_i)}{\lambda_{\max}} \tag{2.1}$$

The retained points form a realisation of events from an inhomogeneous Poisson process with intensity function $\lambda(\boldsymbol{x})$.

2.4.4 Estimating Intensity

Estimation of many point process functions rely on the estimation of the intensity of a stationary point process. Given a sampling region W, a natural unbiased estimator for the intensity of a stationary point process is $\hat{\lambda} = X(W)/\nu(W)$. The intensity function $\lambda(.)$ of an inhomogeneous Poisson process can also be estimated using parametric or non-parametric methods such as kernel-based estimation (see, for example, section 8.5.1 of Cressie 1991).

2.4.5 *Edge-Effects*

Estimating point process functions of interest in some bounded region W, the *sampling* or *observation window*, is not trouble-free. Problems generally encountered are those arising from *edge effects*, that is, estimation problems created by not being able to observe data outside the edges of the observation region. These problems are usually encountered when the region W on which the point pattern is observed is a subset of a larger region on which the process is defined. Therefore estimation of summary statistics is biased by having censored events which may be interacting with events in the observation window. The methods of dealing with edge effects can be split into three categories, the simplest being the use of border methods (Diggle 2003), using estimators that explicitly account for edge effects, and wrapping W into a torus by identifying opposite edges. However, the toroidal wrapping technique does not generally apply to the confocal microscopy data.

2.5 Testing for Spatial Point Processes

2.5.1 *The Empty Space Function F*

Let X be a stationary and isotropic point process. That is, all probability distributions associated with X are invariant under rotation and translation (Baddeley et al. 1992). The *empty space function* of X, denoted F, or $F(r)$ from now onwards, is the probability distribution of the distance from an arbitrary point to the nearest event. That is, for $r \geq 0$

$$F(r) = \mathbb{P}(D(0, X) \leq r) = \mathbb{P}(X(\mathcal{B}^d(0, r)) > 0)$$

where $D(\boldsymbol{x}, A) = \inf\left\{\|\boldsymbol{x} - \boldsymbol{x}'\| : \boldsymbol{x}' \in A\right\}$ is the shortest (Euclidean) distance from $\boldsymbol{x}$ to A. For homogeneous Poisson process with intensity λ is given by

$$F(r) = 1 - \exp\left(-\lambda \nu(\mathcal{B}^d(0, r))\right) = \begin{cases} 1 - \exp\left(-\lambda \pi r^2\right) & d = 2 \\ 1 - \exp\left(-4\lambda \pi r^3 / 3\right) & d = 3 \end{cases} \tag{2.2}$$

Using (2.2) and by estimating the empty space function of a point pattern we can assess whether there is regularity or aggregation (clustering) in a point pattern. Estimated values of $F(r)$ greater than that given by (2.2) suggests that there is regularity, while lower values suggest aggregation (Baddeley et al. 1992).

Baddeley et al. (1992) state that the empty space function is typically estimated by taking a fine grid in the sampling region W and computing the distance from each grid point to the nearest event. This technique results in edge effects as we are unable to search for points outside W. The only approach currently in use is the *border method* (Baddeley et al. 1992). When adopting this technique, only events that are at least a distance r from the boundary of W are considered.

2.5.2 The Nearest Neighbour Distribution Function G

The G function is the distribution of the distance from a *typical event* of the process to the nearest other point of the process. For stationary point process X, the $G(r)$ function associated with X is given by

$$G(r) \quad = \quad \mathbb{P}(D(0, X \setminus \{0\}) \leq r \mid 0 \in X) = \mathbb{P}(X(\mathcal{B}^d(0,r)) > 1 \mid 0 \in X) \qquad r \geq 0$$

where $X \setminus \{0\}$ is the process excluding a point at zero. By stationarity the point 0 can be replaced by any arbitrary point $\boldsymbol{x}$. An alternative definition of the $G(r)$ function using the Campbell-Mecke theorem (see section 4.4 of Stoyan et al. 1995) is

$$G(r) = \frac{\mathbb{E}\left[\sum_{\boldsymbol{x} \in X \cap W} \mathbb{I}_{(0,r]}(D(\boldsymbol{x}, X \setminus \{\boldsymbol{x}\}))\right]}{\mathbb{E}[X(W)]}$$

For a homogeneous Poisson process with intensity λ the $G(r)$ function is given by

$$G(r) = 1 - \exp\left(-\lambda \nu(\mathcal{B}^d(0,r))\right) = F(r) = \begin{cases} 1 - \exp\left(-\lambda \pi r^2\right) & d = 2 \\ 1 - \exp\left(-4\lambda \pi r^3 / 3\right) & d = 3 \end{cases}$$

A border-corrected estimate for the G function is given by

$$\hat{G}(r) = \frac{\sum_{\boldsymbol{x} \in X} \mathbb{I}_{(0,r]}(r(\boldsymbol{x})) \mathbb{I}_{W_{\ominus r(\boldsymbol{x})}}(\boldsymbol{x})}{\sum_{\boldsymbol{x} \in X} \mathbb{I}_{W_{\ominus r(\boldsymbol{x})}}(\boldsymbol{x})}$$

where $r(\boldsymbol{x}) = D(\boldsymbol{x}, X \setminus \{\boldsymbol{x}\})$, and $W_{\ominus r}$ is an *erosion* of W, that is, $W_{\ominus r} = \{\boldsymbol{x} \in W: B^d(\boldsymbol{x}, r) \subset W\}$.

2.5.3 The Pair Correlation Function g

The *pair correlation function*, $g(r)$ is the frequency of event pairs within distance r. The *pair correlation function* is widely used in spatial statistics and particularly in astronomy and astrophysics, for example (Kerscher 1998). Provided that the second order product density exists, then in the stationary and isotropic case we can write the correlation function $\rho^{(2)}(\boldsymbol{x},\boldsymbol{x}')\equiv\rho^{(2)}(r)$ for $r=\|\boldsymbol{x}-\boldsymbol{x}'\|$. The *pair correlation function* is defined for a stationary point process with intensity λ by

$$g(r)=\frac{\rho^{(2)}(r)}{\lambda^2} \tag{2.3}$$

For a Poisson process, we have $g(r)=1$. Furthermore, $g(r)>1$ indicates clustering while $g(r)<1$ is a sign of regularity. The pair correlation function can be estimated using estimator $\hat{\rho}^{(2)}(r)$ for the second order product density.

2.5.4 The K Function

The K function appears at present to be the most popular second order characteristic used in point process analysis. For a stationary point process with intensity λ, $\lambda K(r)$ is the mean number of events that are within distance r of the typical event,

$$K(r)=\frac{\mathbb{E}[X(\mathcal{B}^d(0,r))]}{\lambda}. \tag{2.4}$$

The Campbell-Mecke theorem yields the alternative definition

$$K(r)=\frac{\mathbb{E}\left[\sum_{x\in X\cap B}X(\mathcal{B}^d(\boldsymbol{x},r)\setminus\{\boldsymbol{x}\})\right]}{\mathbb{E}[X(B)]} \tag{2.5}$$

for arbitrary B with $0<\nu(B)<\infty$, where $X(\mathcal{B}^d(\boldsymbol{x},r)\setminus\{\boldsymbol{x}\})$ is the count of the number of points in the ball radius r centered at $\boldsymbol{x}$, excluding $\boldsymbol{x}$. For a homogeneous Poisson process with intensity λ, $K(r)$ is given by

$$K(r)=\nu(B^d(0,r))=\begin{cases}\pi r^2 & d=2\\ 4\pi r^3/3 & d=3\end{cases}$$

A border-corrected estimate of $K(r)$ for region W is

$$\hat{K}(r)=\frac{\nu(W_{\ominus r})}{X(W_{\ominus r})^2}\sum_{x\in W_{\ominus r}}\sum_{x'\in W}\mathbb{I}_{(0,r]}\left(\|\boldsymbol{x}-\boldsymbol{x}'\|\right).$$

2.5.5 Relationships Between Spatial Point Process Functions

The K and g functions are closely related, as K can be expressed in terms of g by the equation

$$K(r)=c(d)\int_0^r u^{d-1}g(u)du \tag{2.6}$$

for some specified constant $c(d)$. Some other characteristics have been defined as combinations and variants of those discussed. Of particular importance is the J-function, suggested by Van Lieshout and Baddeley (1999). For a stationary point process the $J(r)$ function is defined as

$$J(r) = \frac{1 - G(r)}{1 - F(r)}$$

for $F(r) < 1$. The $J(r)$ function is $J(r) = 1$ for a homogeneous Poisson process. However, $J(r) = 1$ does not imply that the point process is a homogeneous Poisson. $J(r)$ can be estimated by using

$$\hat{J}(r) = \frac{1 - \hat{G}(r)}{1 - \hat{F}(r)}$$

In general, for $r > 0$, $J(r) < 1$ indicates clustering and $J(r) > 1$ is a sign of regularity.

An alternative to the K-function is the L-function, defined as

$$L(r) = \left(\frac{K(r)}{\nu(@^d(0,1))} \right)^{1/d}.$$

which can be estimated using the estimate $\hat{K}(r)$. For a homogeneous Poisson process, $L(r) = r$, so that $L(r) - r = 0$.

The pair correlation and K-functions can be defined for the non-stationary case, see Møller and Waagepetersen (2004). The anisotropic versions of these functions are defined by Stoyan and Stoyan (1994). Baddeley et al. (2000) propose definitions for the non-stationary versions of the F and G function in their concluding discussions on the analysis of inhomogeneous point patterns.

2.6 More Complicated Point Process Models

Earlier we discussed the simplest point process, the homogeneous Poisson process. We can divide the most commonly used and more complicated point process models into three categories, inhomogeneous Poisson models, models for point patterns which exhibit clustering, and models for point patterns which are regular. The exception to this classification are Cox processes, an important class of models that can be used to model both clustering and regularity (see chapter 5 of Møller and Waagepetersen (2004)).

Preliminary analysis on PPDS1 provided some possible evidence for clustering and hence our discussions here are favoured towards models for clustered data. The cluster models we discuss briefly include the Matérn cluster process (see for example Cressie (1991)). This model has, for example, been used for modelling tree roots data (Fleischer et al. 2006). We also consider the Gauss–Poisson process

(Stoyan et al. 1995). Point process models tend to be generalisations of other point process models; the homogeneous Poisson process is a special case of the inhomogeneous one, which can be generalised to a Cox process.

Models can also be formed by the three fundamental operations discussed in section 5.1 of Stoyan et al. (1995). These operations include superposition, thinning and clustering. In a clustering operation the events of a point process are replaced by clusters of points, X_0. The clusters ($X_0 s$) themselves are spatial point processes. It is common practice to refer to the events as "parents" and the events of the clusters as "daughters". The two cluster processes we discuss here are members of a group of processes called *Neyman–Scott* processes. Neyman–Scott processes result from homogeneous independent clustering applied to a stationary Poisson process. Some Neyman–Scott process such as the Matérn cluster process are also Cox processes.

2.6.1 Gauss–Poisson Process

A *Gauss–Poisson* process (Newman 1970) is an example of a Poisson cluster process (Stoyan et al. 1995). The parent points have a homogeneous Poisson distribution with intensity λ and the number of daughters of each parent is one, two or three with probability q_0,q_1, and q_2 respectively. If the parent has one daughter then the daughter is placed at the parent location. If the parent has two daughters then one is placed at the parent and the other is placed randomly at distance s from the first daughter. The resulting pattern only includes daughter points (and hence the parent points are deleted). Some further results for Gauss–Poisson processes can be found in Milne and Westcott (1972).

2.6.2 Matérn Cluster Process

Matérn's cluster process consists of parents that come from a homogeneous Poisson point process with intensity λ_p. Each parent has m daughters which are uniformly distributed inside $\mathcal{B}^d(0, R)$ (with the parent point being regarded as the origin). The parameter m comes from a Poisson distribution with intensity λ_m. Implicit expressions for the K and g function for a Matérn cluster process can be found in Stoyan et al. (1995).

The Matérn cluster process and Gauss–Poisson process can be simulated in the compact window W directly via the model definitions. Although care should be taken with regards to edge effects. A simple way to account for edge effects is to simulate the parent points inside the dilated window $W_{\ominus \mathcal{B}^d(0,R)}$ where R is such that for the $P(X_0 \supset \mathcal{B}^d(0,R))$ is very small or zero (Stoyan et al. 1995). Brix and Kendall (2002) discuss the simulation of cluster point processes without edge effects.

2.6.3 Markov Point Processes

Markov or Gibbs point processes have been intensively used in spatial statistics since 1970 (Stoyan and Stoyan 1994). Although they are models for various types of point patterns, they are usually recognised for their ability to provide a more flexible framework for modeling spatial point patterns that exhibit inhibition (compared to a homogeneous Poisson distribution) (Cressie 1991). Markov point processes were first defined by Ripley and Kelly (1977). As redefined by Cressie (1991), a spatial point process on bounded set $V \subset \mathbb{R}^d$ is said to be *Markov of range* ρ if it is a spatial point process that has conditional intensity at $\boldsymbol{x} \in V$ given the realisation of the process in $A \backslash \boldsymbol{x}$ that depends only only on the events in $\mathcal{B}^d(\boldsymbol{x}, \rho) \setminus \boldsymbol{x}$. Each Markov process is characterised by a likelihood ratio $f(.)$ with respect to a unit intensity Poisson process. Furthermore, $f(.)$ is usually defined up to a normalising constant that cannot be evaluated in closed form Diggle (2003). A popular example of a Markov point process is the Strauss process (Strauss 1975). In this case, for a configuration of $n < \infty$ points, we have

$$f(x) = \alpha \beta^n \gamma^{\varphi(R)}, \quad \beta > 0, 0 \leq \gamma \leq 1, \; R > 0$$

Where $\varphi(R)$ is the number of distinct pair of events within distance r. The *Papangelou conditional intensity* defined by

$$\lambda^*(\boldsymbol{x}, \boldsymbol{x}') = \frac{f(\boldsymbol{x} \cap \boldsymbol{x}')}{f(\boldsymbol{x})}, \quad \boldsymbol{x}' \in V \setminus \boldsymbol{x}$$

where we take $a/0 = 0$ for $a \geq 0$ (Kallenberg 1984) is a fundamental characteristic (Møller and Waagepetersen 2001). If f is hereditary (that is $f(x) > 0 \Rightarrow f(y) > 0$ for $y \subset x$), then there is a one-to-one correspondence between f and λ^*. Distribution characteristics (such as the summary statistics introduced earlier) for Markov models are difficult to calculate (Stoyan and Stoyan 1994). Further theory on Markov point process can be found in Stoyan et al. (1995) while a good exposition on simulating Markov point processes can be found in Møller and Waagepetersen (2001) and Møller and Waagepetersen (2004).

2.6.4 Cox Processes

A Cox process is a natural approach for generalising the definition of Poisson point process (Møller and Waagepetersen 2004). A Cox Process on $V \subset \mathbb{R}^d$ is often referred to as a 'doubly stochastic' Poisson point process as the intensity measure is replaced by a random locally finite measure Z_Λ. More formally, we say that a point process X is a Cox process driven by Z_Λ if $X | Z_\Lambda = \Lambda$ is an inhomogeneous Poisson process with mean measure Λ. Due to their generality and associated manageable closed form calculations, Cox processes tend to find important applications as stochastic models (Stoyan et al. 1995). Examples of Cox processes include the Matérn cluster process. A particular useful class of Cox processes is the class of *Log Gaussian Cox Processes*. A detailed account of Cox processes can be found in Møller and Waagepetersen (2002).

2.7 Marked Spatial Point Processes

A rigorous definition of a point process can be found in Karr (1991). A *marked spatial point process* is a mathematical model for random or irregularly placed points lying in some two- or three-dimensional region, for which each point realization has an associated *mark*, a random variable representing the magnitude or type of some feature that can be measured at that spatial location. A *multivariate spatial point pattern* is a special case of a marked spatial point pattern, where there is a finite number of marks, each representing an event-type (Cressie 1991). A bivariate spatial point process may be used to model the locations of two different types of subnuclear bodies in the nucleus, while a marked spatial point process may be used (as done in this paper) to model the size of one type of subnuclear body in the nucleus.

Spatial pattern analysis (whether marked or unmarked) often begins with tests to determine whether objects are uniformly placed within a specified region. For PML data this is equivalent to testing whether the PML NBs are randomly placed within the cell nucleus. Several techniques have been adopted for assessing the spatial distribution of nuclear bodies. A popular and relatively fast approach for assessing whether nuclear bodies exhibit spatial positioning preference is known as erosion or "nuclear peeling" (Shiels et al. 2007); this entails some form of radial analysis, in which the nucleus is subdivided into concentric rings or shells from the periphery to the centre. Other techniques for investigating subnuclear body spatial preference have included those adopted by Bolzer et al. (2005). They used the mean of inter-body distances and Kolmogorov–Smirnov tests to assess the spatial distribution of subnuclear bodies.

Tests for uniformity are known as tests for CSR (see Section 2.4.2). Such tests will often provide useful insight into any spatial features (such as clustering) and they often entail computing and interpreting (possibly several) distance-based summary statistics. Estimation of these statistics is generally complicated by edge effects. This issue arises for the PML data because the cell nucleus is assumed to cover a finite bounded region and thus estimation of the statistic (which are potentially defined for unbounded regions) can be biased. CSR tests performed on PML data PPDS1, using estimates of the $F(r)$-function (which, informally, is the probability that a PML NB is within distance r of an arbitrary chosen other PML NB), provided some evidence to reject the null hypothesis of CSR (see Umande 2008). Thus there is evidence that PML are not uniformly placed inside the nucleus.

2.8 Bivariate Spatial Point Process Analysis of PML NBs and RNA Polymerase II

One of the most popular summary statistics used for CSR tests in the univariate case is the $K(r)$-function (Diggle 2003). The bivariate version of the $K(r)$-function, $K_{ij}(r)$ of a stationary (invariant under translation) marked point process was first introduced by Hanisch and Stoyan (1979). Heuristically, letting λ_k denote the intensity of events of type X_k, $K_{ij}(r)$ is the expected number of events of type j that are within distance r of an event of type i. Informally, this means, if X_i denotes the location of PML NBs and X_j RNA Polymerase II then λ_i is the average number of PML NBs per unit volume of the cell nucleus and $K_{ij}(r)$ is the average number of PML NBs that are within a distance r of the RNA Polymerase II.

For n_1 the number of type 1 events, and n_2 the number of type 2 events, the $K_{ij}(r)$-function can be estimated inside the bounded window $W \subset \mathbb{R}^d$ (i.e. the interior of the nucleus) using the estimator

$$\widehat{K}_{ij}(r) = \frac{\sum_{x \in X_1} \sum_{x' \in X_2} \omega(x, x')^{-1} \mathbb{I}_{(\|x - x'\|) \le r}}{\nu(W)\hat{\lambda}_1 \hat{\lambda}_2} \tag{2.7}$$

as suggested by Hanisch and Stoyan (1979). Here $\nu(W)$ is the volume of the nuclear interior. The estimate for the intensity parameter, $\hat{\lambda}_i$ is given by $\hat{\lambda}_i = n_i / \nu(W)$. The function ω is an edge-correction factor such as the proportion of the surface area of the three-dimensional ball centred at $\boldsymbol{x}$, passing through $\boldsymbol{x}'$. For relative ease of calculation and efficiency, we prefer the edge-correction $\omega(\boldsymbol{x},\boldsymbol{x}') = \nu(W \cap b(\boldsymbol{x}, \|\boldsymbol{x} - \boldsymbol{x}'\|))$. To adopt this preferred form of edge-correction, we can utilise the quadrature approximation

$$\nu\left(W \cap b\left(\boldsymbol{x}, \|\boldsymbol{x} \quad \boldsymbol{x}'\|\right)\right) \approx \frac{\nu(W)}{U} \sum_{i=1}^{U} \mathbb{I}_{\left(x_i \in W \cap b\left(x, \|x - x'\|\right)\right)} \tag{2.8}$$

We estimated the $K_{12}(r)$-function using (2.7) and the preferred form of edge-correction (2.8) for cell 4 of PPDS3. The data used to produced Fig. 2.3 suggest that, heuristically, for this particular cell, we would expect to observe fewer RNA Polymerase II bodies within a distance of 0.5 units of the typical PML, compared to if the RNA Polymerase II exhibited CSR. From the biological literature, it appears that the inhibition (and perhaps more generally, spatial relationship between RNA Polymerase II and PML NBs) is driven by the biological function of PML NBs with different cell nuclei. Also, it may be interesting to compare these results with the findings of Xie and Pombo (2006) who reported that PML bodies contain no detectable RNA polymerase II, but are often surrounded by them at a distance greater than 25 nm. From the detail provided in the Appendix, we estimate that 0.5 units is approximately 41 nm.

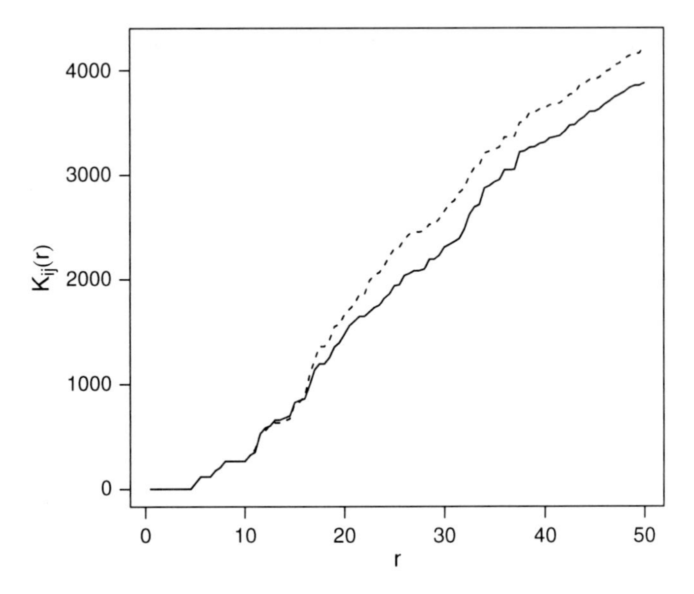

Fig. 2.3 The $K_{12}(r)$-function plot for the PML NB (type 1) and RNA Polymerase II (type 2) in cell nucleus 4 of PPDS3. The solid curve is K_{12} and the dashed curve is K_{21}

Following investigation on edge-correction provided by Umande (2008), we are cautious towards interpreting the results provided in Fig. 2.3 for large values of r. We therefore also consider the outcome of simulation studies. Specifically, a plot of $K_{12}(r)$-functions with simulation envelopes, as shown in Fig. 2.8, can provide further insight. The $K_{12}(r)$-function simulation envelopes for cell nucleus k of PPDS3 was obtained by simulating 100 independent realisations of a homogeneous Poisson process, $\tilde{x}_{1k}, \ldots, \tilde{x}_{100k}$, inside the convex hull, representing the nuclear boundary of cell k. Each simulated realisation, $\tilde{x}_{jk}$, of the homogeneous Poisson process had a conditional number of events, n_k, where n_k is the number of RNA Polymerase II found inside cell nucleus k. For each $\tilde{x}_{jk}$ we estimate the $K_{12}(r)$-function without applying an edge-correction, where 1 is an event of type PML and 2 is an event belonging to $\tilde{x}_{jk}$. As we are conducting a like-for-like comparison, in the sense that a biased estimate of the $K(r)$-function is being compared to a another biased estimate of the same function, there is no need for an edge-correction. We hence obtain 100 estimates of the $K_{12}(r)$-function, $\{K_{12_{1k}}(r), \ldots, K_{12_{100k}}(r)\}$ for each cell k.

The upper and lower simulation envelopes for each cell nucleus are respectively $\inf\{K_{12_{1k}}(r), \ldots, K_{12_{100k}}(r)\}$ and $\sup\{K_{12_{1k}}(r), \ldots, K_{12_{100k}}(r)\}$. The results presented in Fig. 2.8 suggest that, apart from cell 4, generally, for a wide range of r, there are fewer RNA Polymerase II bodies within distance r of the typical PML body compared to RNA Polymerase II bodies randomly scattered inside the nucleus. The results for cell 4 are consistent with those obtained for small r (relative to the nuclear magnitude), for the other cells in PPDS3. However, on the contrary to the other cells, for larger r, the PML in cell nucleus 4 typically tend to have a much greater number of RNA Polymerase II bodies within distance r, compared to RNA

Polymerase II randomly placed inside the nucleus. Simulation studies also confirmed that there are enough nuclear bodies in cell 4 of PPDS3 for one to make legitimate observations for features of the spatial point pattern at distances of 0.5 units. This is owed to the high number of RNA Polymerase II bodies (there are 72 RNA Polymerase II bodies in cell 4 of PPDS3).

2.9 A Marked Inhomogeneous Poisson Process Model for PML

Obtaining a model for the spatial distribution of PML NBs is important. Below, we outline how one could fit an inhomogeneous Poisson process to the type of PML data used throughout this paper. By successfully fitting an inhomogeneous Poisson process to replicated PML NB data, suggests a form of a spatial preference for PML within the cell nucleus. Also, very importantly, the formulation of an appropriate model can have potential applications in spotting certain illnesses by comparing the distribution of PML NBs from cells that have been taken from the subject being diagnosed, with the distribution of PML NBs as suggested by the model. At present, this is of course rather ambitious, given the technological limitations.

We have hinted above at a possible candidate model to describe the spatial locations of PML NBs. On rejecting the null hypothesis of CSR, a model that is commonly considered is the inhomogeneous Poisson process. The inhomogeneous Poisson process model is essentially the model that would generate CSR data but with a spatially varying parameter $\lambda(.)$. In terms of the PML NB data, under this model, the number of PML NBs per unit volume is assumed to vary throughout the nuclear interior. The inhomogeneous Poisson process intensity function $\lambda(\boldsymbol{x})$ determines how the PML NBs are distributed throughout the nuclear interior; determining $\lambda(\boldsymbol{x})$ is the core modelling challenge.

Practitioners might consider biological literature when attempting to specify $\lambda(\boldsymbol{x})$. For example, McManus et al. (2006) have reported that chromosomes and regions of chromosomes segregate differently within the nucleus depending on whether or not they are rich in potentially transcribed genes. The individual interphase chromosome territories segregate their gene rich R-bands into the interior of the nucleoplasm, whereas their gene poor G-bands are gathered against the periphery of the nucleus and against the nucleolar surface (see for example Shopland et al. (2003)). Euchromatin sequences are further organised such that they maintain a spatial relationship with the predominant nucleoplasmic nonchromatin structure, the splicing factor compartments (McManus et al. 2006). Smaller nonchromatin structures such as PML associate with specific regions of the genome. In summary, this means there is biological reason to suggest that the spatial location of PML NBs is related to the nuclear boundary. Umande (2008) has used simulation studies and a variant of the empty space function to determine a possible relationship between the placement of PML NBs and the nuclear boundary.

A candidate model that stems from these ideas is one defined through the following postulates:

- MP1 The event (PML NB) locations are a realisation of a homogeneous Poisson process with intensity λ inside bounded $W \subset \mathbb{R}^3$
- MP2 Each event x is retained with probability

$$p(\boldsymbol{x}) = 1 - \exp\left(-\kappa \left\|\boldsymbol{x} - \partial W\right\|\right) \kappa \in \mathbb{R}^+$$

Otherwise independently thinned (removed) with probability $1 - p(\boldsymbol{x})$ where ∂W denotes the boundary of W.

A model defined through postulates MP1-MP2 is an inhomogeneous Poisson process with intensity function $\lambda p(x)$ (see Umande (2008) for a mathematical proof). Furthermore, note that under this model, as $\| \boldsymbol{x} - \partial W \| \to 0$, $p(\boldsymbol{x}) \to 0$ which means that PML NBs are less likely to be observed close to the boundary.

We can fit Model 1 to PPDS2 as follows. We first note that for a single replicate, the likelihood, l, of the data $\mathcal{D}$, is given by

$$\ell = p(\mathcal{D}) p(M \mid \mathcal{D}).$$

For k iid replicates the likelihood l_{Rep} is given by

$$\ell_{\text{Rep}} = \prod_{j=1}^{k} p(\mathcal{D}_j) p(\mathcal{M} \mid \mathcal{D}_j)$$

and the log-likelihood is given by

$$\ell = \sum_{j=1}^{k} \log(p(\mathcal{D}_j)) + \log(p(\mathcal{M} \mid \mathcal{D}_j))$$

We can therefore fit the marks separately to the model. However, note that before modelling replicated data that one is uncertain follow the same statistical distribution, it is advisable to begin by testing whether or not the data is "similar" (i.e. whether the data truly does come from the same statistical distribution). Diggle (2003) and Webster et al. (2006) provide details on tests that can be used for testing spatial point pattern similarity. The procedures are not straightforward when applied to data analysed here; bootstrapping techniques and a non-stationary version of the $K(r)$-function are used.

We will now provide an exposition on how we can mark the PML NBs and gain initial insight into the mark distribution by analysing an appropriate spatial point process characteristic and can thus use the marks analysis to completely specify a marked point process model for the PML NB spatial locations. As mentioned above, the extension of other popular characteristics to the multivariate case is generally not difficult (for discrete marks). For the general marked case, the empty space function, $F(r)$ of the marked spatial point process $X^{[m]}$ is the cumulative distribution function of the distance from a randomly selected origin to the nearest event in $X^{[m]}$. That is

$$F(r) = P\left(X^{[m]} \cap (b(0,r) \times \mathbb{M}) \neq \emptyset\right)$$

Also, let B be a subset of $\mathbb{M}$ with $Z_{X^{[m]}}(B) > 0$. We define the nearest neighbour function for events with marks in B by

$$G_B(r) = P^!_{X^{[m]},0}\left(X^{[m]} \cap (b(0,r) \times \mathbb{M}) \neq \emptyset\right)$$

for $r \geq 0$. Here $P^!$ denotes a probability with respect to the Palm distribution. Van Lieshout (2004) introduced a *J*-function for marked spatial point patterns. The *J*-function with respect to mark set *B*, J_B is given by

$$J_B(t) = \frac{1 - G_B(t)}{1 - F(t)}$$

for all $t \geq 0$ and $F(t) < 1$. For an independently marked Poisson process, $G_B(t) = F(t)$ for all *t* and so $J_B \equiv 1$. Values greater than 1 are a sign of inhibition, while values less than 1 are a sign of clustering.

Van Lieshout (2004) proved, for *X* a stationary point process on $\mathbb{R}^d$ with intensity $0 < \lambda < \infty$, that if *X* is *randomly labelled* with mark distribution *Z* on mark space $\mathbb{M}$ and if $X^{[m]}$ is the marked point process obtained, then for all $r \geq 0$ with $F(r) < 1$, the *J*-function with respect to a mark set $B \subset \mathbb{M}$ with $Z_{X^{[m]}}(B) > 0$ is given by

$$J_B(r) = J_X(r)$$

where $J_X(r)$ is the *J*-function of *X* and where the marked spatial point process $X^{[m]}$ is said to have the *random labelling* property if the marks of the events are conditionally iid given the event locations.

For each PML NB in PPDS2, we calculated an approximate PML body length from the image data used to produce PPDS2. This was done by measuring the maximum distance between any two points, of the points that have been classified as being a part of that PML NB in the image processing stage. That is, in the data provided by the Imperial College London centre for structural biology, each PML NB *j* is described as a set of points $\{x_{1j}, \ldots. x_{uj}\}$ (see Appendix). The length of PML NB *j* was calculated as $\inf\left\{\left\| x_{ij} - x_{sj} \right\| : i, s = 1, \ldots., u\right\}$. We use these lengths to assign marks to the PML NBs in PPDS2.

The *J*-function plots for the cell nuclei of PPDS2 is shown in Fig. 2.4. Figures 2.4 and 2.5 suggests that, since the marked and unmarked *J*(*r*)-functions are not too dissimilar, we would generally not necessarily expect to observe the PML NBs placed in the nuclear interior, in such a way that depends on their relative sizes (in terms of length). Note also that we found that the proportions of the PML body length to nuclear length, denoted by z_π was consistent with the theoretical proportions provided in the biological literature. All of the PML NBs in PPDS2 were pooled and we calculated the linear correlation between z_π and the proportion of PML NB distance to the boundary to nuclear length. We obtained a correlation of 0.09, suggesting that the two are not strongly linearly correlated. The lack of correlation between the length of the PML and distance to boundary, provides some evidence for random labelling with respect to PML size. This is consistent with the results obtained using the marked *J*-function. Hence, these results would not support for example, a view that larger PML NBs are found closer to the nuclear periphery or more internally.

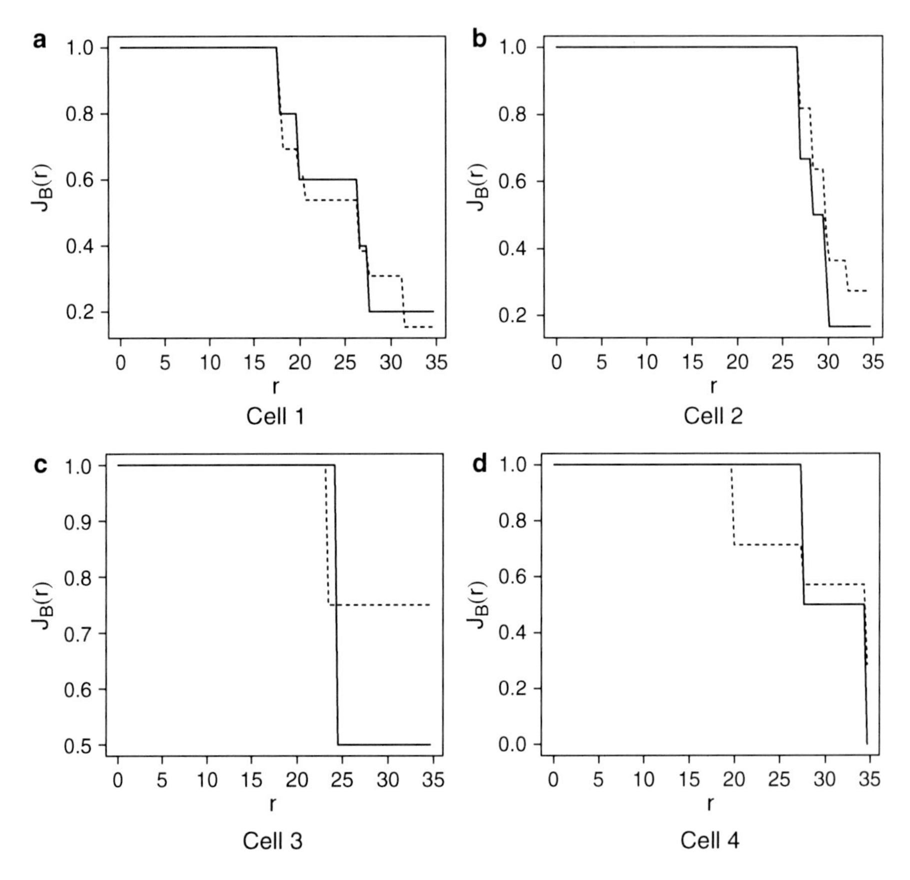

Fig. 2.4 Marked (*solid curve*) and unmarked (*dashed line*) $J(r)$-function for PPDS2 cells 1–4. The mark set $B = [0.01, 0.04]$

The final step in the model fitting discussions endeavours to identify each PML NB uniquely by assigning a mark to the PML NB. Each PML NB is now assigned a length. Consider the additional model postulate:

- MP3 Each PML NB $\boldsymbol{x}$ is randomly assigned a proportional length $z_\pi \sim Z$. That is, Z is a random variable that assigns to each PML NB, the mark

$$z_\pi = \frac{\text{PML NB length}}{\text{nuclear length}}.$$

Formal tests on the data (see Fig. 2.6) suggest that a normal distribution is a plausible model for the marks distribution Z.

A Kolmogorov–Smirnov test for the null hypothesis that the marks follow a normal distribution with mean 0.045 and standard deviation 0.019 provided a p-value of 0.92. Caution is required when choosing the mark space since physical restrictions mean that, realistically, the mark space (that the z_π belong to) is $A \subset (0, 1)$

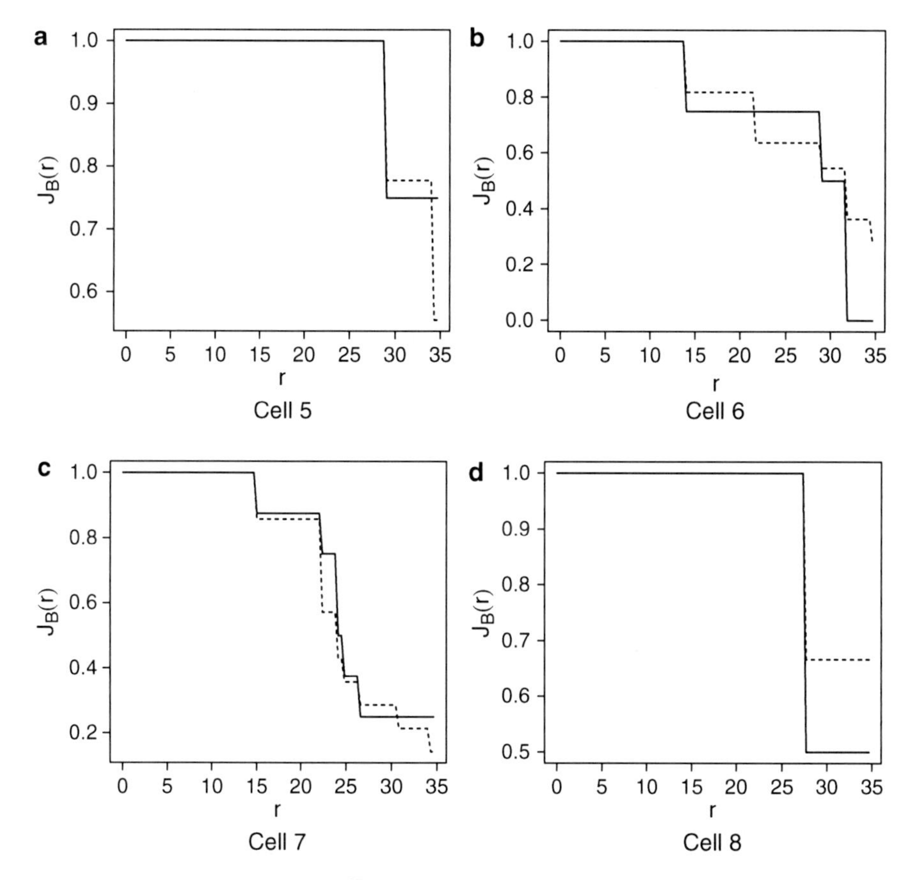

Fig. 2.5 Marked (*solid curve*) and unmarked (*dashed line*) $J(r)$-function for PPDS2 cells 5-8. The mark set $B=[0.01, 0.04]$

(and not for example $\mathbb{R}^+$ as implied by a normal distribution). This is because the PML NBs cannot be longer than the nucleus or have zero length. Hence, it is more appropriate to adopt a truncated normal distribution for Z. More precisely, Z has a normal distribution and lies within the interval (0,1). We estimated the mean and variance of the truncated (0,1) normal distribution, for the PML NB marks, to being (respectively) 0.045 and 0.019 (see for example Barr and Sherrill (1999) for detail on the parameter estimation). The diagnostic plots presented in Fig. 2.7 suggest that the truncated normal model that has been put forward for the PML NB marks distribution is a plausible one.

By using this model for the marks distribution as Z in MP3, and by letting MP3 be an additional final postulate of the Model defined by MP1-MP2, we obtain a marked spatial point process model for the spatial distribution of PML NBs. We may also wish to assess how well the inhomogeneous Poisson process model fits the data. Umande (2008) has carried out such tests on data similar to that used in

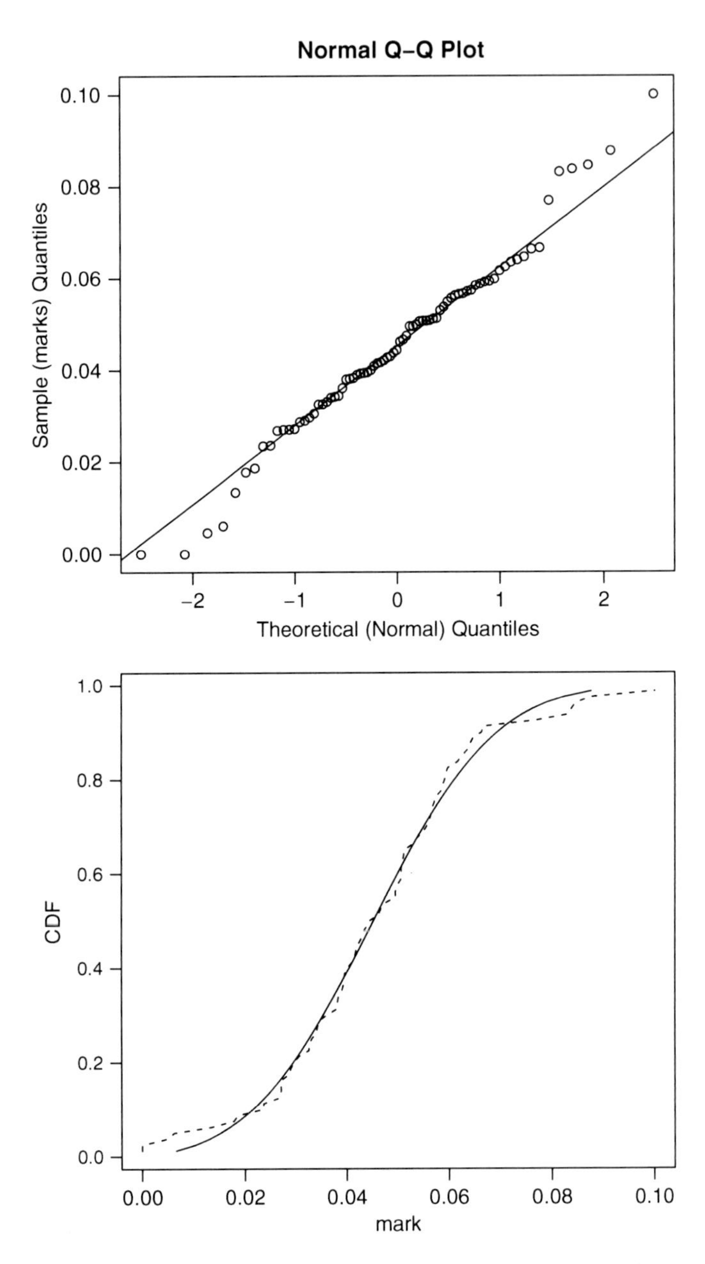

Fig. 2.6 Q–Q plot of the PML NB marks (*top*). The points are approximately linear, suggesting possible normality. The graph on the bottom shows the empirical CDF of the PML NB marks (*dashed curve*) with the CDF of a normal distribution with mean 0.045 and standard deviation 0.019 (*black line*)

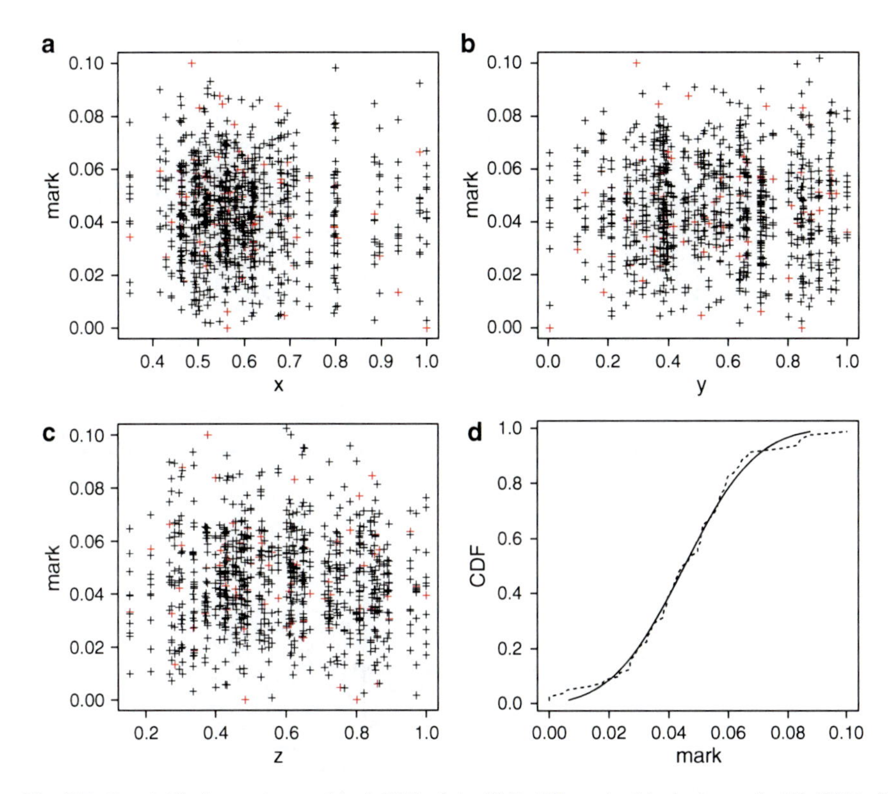

Fig. 2.7 Panel (**d**) shows the empirical CDF of the PML NB marks (*dashed curve*) with CDF of a truncated normal distribution with mean 0.045 and standard deviation 0.019 (*black line*). The scatter plots are of PML NB centroid (x,y,z) coordinates (divided by nuclear length) against simulated realisations of a truncated (0,1) normal distribution with mean 0.045 and standard deviation 0.019 (the model marks distribution). The *red crosses* represent the data and the *black crosses* are for the simulated marks

this paper and found a model defined through MP1-MP2 as a credible model for PML NB locations.

2.10 Conclusion

Tools from spatial point pattern analysis can be invaluable in the investigation of the configuration of nuclear bodies, in particular, the way that PML bodies are distributed across the nucleus in relation to themselves and to other nuclear bodies. By computing inter-object distances and the corresponding K and J functions, simulation-based statistical tests of hypotheses can be formulated and implemented, and these tests allow the validity of important biological models to be assessed.

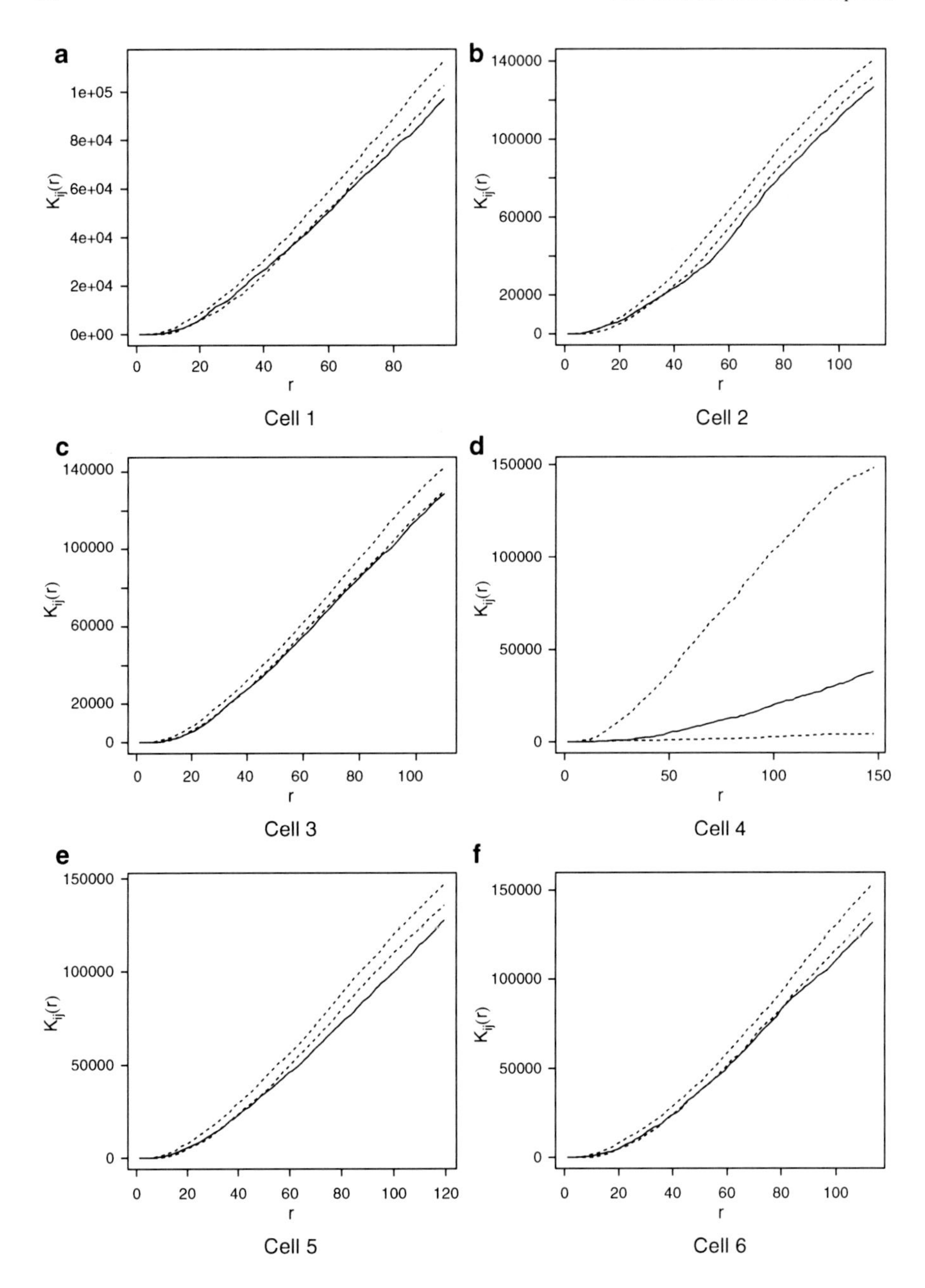

Fig. 2.8 The $K_{12}(r)$-function plot for the PML NB (type 1) and RNA Polymerase II (type 2) in cell nuclei 1–6

Acknowledgements The authors would like to thank Dr. Niall Adams and Professor Paul Freemont for their substantial involvement in this work, and Dr. Elizabeth Batty and Dr Carol Shiels for their experimental work. David Stephens is supported by a Natural Sciences and Engineering Research Council of Canada (NSERC) Discovery Grant.

Appendix

All of the datsets refered to in this paper (PPDS1, PPDS2 and PPDS3) were provided by the Imperial College London Centre for Structural Biology. The cells used are MRC-5 cell nuclei. Furthermore, the cells used for PPDS2 are all in the G_0 phase of the cell cycle.

- **Distance Units for the Data** Throughout this paper we refer to "units" for reporting measured distances in the cell nuclei. r units corresponds to r image pixels. There are 12 pixels in 1 micrometer (μm). Hence 1 unit$\approx$0.083 μm or 83.3 nanometers (nm).
- **PPDS1** The dataset PPDS1 consists of five cells. We are provided with point coordinates that are the centroids of PML NBs inside the cell nuclei. The sampling region is obtained by calculating the smallest ellipsoid that contains all of the PML points.
- **PPDS2 and PPDS3** Once the confocal image is produced, PPDS2 is obtained by analysing the image data, to form a datset consisting of points that are labeled PML, nucleoli, nuclear boundary, or empty space inside the nucleus. We then run this (large) dataset trough a computer program that forms a point pattern by converting the PML NBs into PML points, by calculating their centroids. PPDS3 is produced in a similar way to PPDS2 but contains an additional labelling to indicate the locations of RNA Polymerase II. Further details, including the number of interior points, U, of PPDS2 and PPDS3 are shown in Tables 2.1 and 2.2.

Table 2.1 PPDS2 details

Cell	PML Count	U
1	12	49,942
2	11	50,189
3	9	49,975
4	7	49,969
5	9	49,900
6	11	49,974
7	14	50,091
8	6	50,078

Table 2.2 PPDS3 details

Cell	PML Count	RNA Pol II Count	U
1	6	349	50,216
2	8	269	50,253
3	13	296	50,266
4	12	72	50,266
5	10	226	50,250
6	14	125	49,982

References

Baddeley AJ, Moyeed RA, Howard CV, Boyde A (1992) Analysis of a three-dimensional point pattern with replication. Appl Statist 42(4):641–668

Baddeley AJ, Kerscher M, Schladitz K, Scott BT (2000) Estimating the J function without edge correction. Stat Neerl 54:315–328

Barr D, Sherrill E (1999) Mean and variance of truncated normal distributions. Am Statist 53(4):357–361

Bolzer A, Kreth G, Solovei I, Koehie D, Fauth C, Muller S, Eils R, Cremer C, Speicher MR, Cremer T (2005) Three-dimensional maps of all chromosomes in human male fibroblast nuclei and prometaphase rosettes. PLOS Biol 3(5):826–842

Borden KLB (2002) Pondering the promyelocytic leukemia protein (PML) puzzle: possible functions for PML nuclear bodies. Mol Cell Biol 22(15):5259–5269

Brix A, Kendall WS (2002) Simulation of cluster point processes without edge effects. Adv Appl Probabil 34(2):267–280

Cox DR, Isham V (1980) Point processes. Chapman and Hall, New York

Cressie N (1991) Statistics for spatial data. Wiley, New York

Dellaire G, Bazett-Jones DP (2004) PML nuclear bodies: dynamic sensors of dna damage and cellular stress. BioEssays 26(9):963–977

Diggle PJ (2003) Statistical analysis of spatial point patterns. Arnold, London

Fishman GS (1996) Monte Carlo. Springer, New York

Fleischer F, Beil M, Kazda M, Schmidt V (2006) Analysis of spatial point patterns in microscopic and macroscopic biological image data. In: Baddeley A, Gregori P, Mateu J, Stoica R, Stoyan D (eds) Case studies in spatial point process modeling, Lecture Notes in Statistics. Springer, Berlin, pp 232–259

Glasbey CA, Roberts IM (1997) Statistical analysis of the distribution of gold particles over antigen sites after immunogold labelling. J Microsc 186(3):258–262

Hanisch KH, Stoyan D (1979) Formulas for the second-order analysis of marked point processes. Statist J Theoret Appl Statist 10(4):555–560

Kallenberg O (1984) An informal guide to the theory of conditioning in point processes. Int Statist Rev 52(2):151–164

Karr AF (1991) Point processes and their statistical inference. Marcel Dekker, New York

Kemp CD (1988) Review: [untitled]. Statistician 37(1):84–85

Kerscher M (1998) Regularity in the distribution of superclusters? Astron Astrophys 336:29–34

Lanctot C, Cheutin T, Cremer M, Cavalli G, Cremer T (2007) Dynamic genome architecture in the nuclear space: regulation of gene expression in three dimensions. Nature 8:104–115

Lewis PAW, Shedler GS (1976) Simulation of nonhomogeneous Poisson processes with log linear rate function. Biometrika 63(3):501–505

McManus KJ, Stephens DA, Adams NM, Islam SA, Freemont PS, Hendzel MJ (2006) The transcriptional regulator CBP has defined spatial association within interphase nuclei. PLOS Computat Biol 2(10):1271–1283

Milne RK, Westcott M (1972) Further results for Gauss-Poisson processes. Adv Appl Probabil 4(1):151–176

Møller M, Waagepetersen RP (2001) Simulation based inference for spatial point processes, URL citeseer.ist.psu.edu/469797.html

Møller J, Waagepetersen RP (2002) Statistical inference for Cox processes. In: Lawson AB, Denison DGT (eds) Spatial cluster modelling. Chapman and Hall, Boca Raton, FL, p 37

Møller J, Waagepetersen RP (2004) Statistical inference and simulation for spatial point processes. Chapman & Hall, Boca Raton

Newman DS (1970) A new family of point processes which are characterised by their second moment properties. J Appl Probabil 7(2):338–358

Ripley BD (1987) Stochastic simulation. Wiley, New York

Ripley BD, Kelly FP (1977) Markov point processes. Lond Math Soc 15:188–192

Shiels C, Adams NM, Islam SA, Stephens DA, Freemont PS (2007) Quantitative analysis of cell nucleus organisation. Computat Biol 3:1161–1168

Shopland LS, Johnson CV, Byron M, McNeil J, Lawrence JB (2003) Clustering of multiple specific genes and gene-rich r-bands around sc-35 domains: evidence for local euchromatic neighbourhoods. J Cell Biol 162:981–990

Skellam JG (1952) Studies in statistical ecology: I. Spatial pattern. Biometrika 39(3/4):346–362

Stoyan D (2006) Fundamentals of point process statistics. In: Baddeley A, Gregori P, Mateu J, Stoica R, Stoyan D (eds) Case studies in spatial point pattern modelling, no. 185 in Lecture Notes in Statistics. Springer, New York, p 3

Stoyan D, Stoyan H (1994) Fractals, random shapes and point fields. John, Chichester

Stoyan D, Kendall W, Mecke J (1995) Stochastic geometry and its applications, 2nd edn. Wiley, Chichester

Strauss DJ (1975) A model for clustering. Biometrika 62(2):467–475

Thompson HR (1955) Spatial point processes, with application to ecology. Biometrika 42(1/2):102–115

Umande P (2008) Spatial point pattern analysis with application to confocal microscopy data. Ph.D. thesis, Imperial College London

Van Lieshout MNM (2004) A J-function for marked point patterns. Inst Statist Math 511–532

Van Lieshout MNM, Baddeley AJ (1999) Indeces of dependence between types in multivariate point patterns. Scand J Statist 26:511–532

Wang J, Shiels C, Sasieni P, Wu PJ, Islam SA, Freemont PS, Sheer D (2004) Promyelocytic leukemia nuclear bodies associate with transcriptionally active genomic regions. J Cell Biol 164(4):515–526

Webster S, Diggle PJ, Clough HE, Green RB, French NP (2006) Strain-typing transmissible spongiform encephalopathies using replicated spatial data. In: Baddeley A, Gregori P, Mateu J, Stoica R, Stoyan D (eds) Case studies in spatial point pattern modelling, no. 185 in Lecture Notes in Statistics.Springer, New York, p 197

Xie SQ, Pombo A (2006) Distribution of different phosphorylated forms of rna polymerase ii in relation to cajal and PML bodies in human cells: an ultrastructural study. Histochemist Cell Biol 125:21–31

Chapter 3
Quantitative Approaches to Nuclear Architecture Analysis and Modelling

Daniel Hübschmann, Nikolaus Kepper, Christoph Cremer, and Gregor Kreth

Abstract The spatial organisation of the genome in the cell nucleus has emerged as a key element to understand gene function. A wealth of molecular and microscopic information has been accumulated, resulting in a variety of – sometimes contradictory – models of nuclear architecture. So far, however, a large part of this structural information and in consequence also the models derived from them are 'qualitative'. In this overview, a brief introduction will be given into quantitative experimental and modelling approaches to large scale nuclear genome architecture in human cells. As a biomedical application example, the use of a quantitative computer model of the 3D architecture allowed to explore different implications of nuclear structure on chromosomal aberrations. In addition, we shall present two novel examples for quantitative computer modelling: (1) The impact of SC 35 splicing domains on nuclear genome structure; (2) The dynamics of large scale nuclear genome structure in a Brownian motion model. Finally, we shall discuss some perspectives to extend quantitative nuclear structure analysis to the nanoscale.

Keywords Nuclear architecture • Computer modelling • Chromatin domains • Dynamics • SC35 domains • Speckles • R-bands • Superresolution • Lightoptical nanoscopy

---

D. Hübschmann and G. Kreth (✉)
Applied Optics and Information Processing, University Heidelberg, Im Neuenheimer Feld, D-69120, Heidelberg, Germany
e-mail: gkreth@kip.uni-heidelberg.de

N. Kepper
Applied Optics and Information Processing and BioQuant Center, University Heidelberg, Im Neuenheimer Feld, D-69120, Heidelberg, Germany

C. Cremer (✉)
Applied Optics and Information Processing; BioQuant Center, Institute for Pharmacy and Molecular Biotechnology, University Heidelberg, Im Neuenheimer Feld, D-69120, Heidelberg, Germany
and
Institute for Molecular Biophysics, The Jackson Laboratory, 04609 Bar Harbor, ME, USA
e-mail: cremer@kip.uni-heidelberg.de

N.M. Adams and P.S. Freemont (eds.), *Advances in Nuclear Architecture*,
DOI 10.1007/978-90-481-9899-3_3, © Springer Science+Business Media B.V. 2011

3.1 Introduction

A cell nucleus contains a hierarchy of connected levels of dynamical order: The first level is formed by the linear sequence of the of the nuclear DNA chains, in human cells about $2 \times 3\ 10^9$ base pairs. The second level is given by direct interactions between specific parts of the DNA chains and other macromolecules in the cell nucleus (Rippe et al. 2008). Specific proteins or protein complexes may bind to specific DNA sites and thus activate or silence specific genes. The third level of nuclear organization is the dynamic three dimensional organization of the DNA chains: For example, in each human lymphocyte, the about 2 m long DNA has to be packaged in such a way that it fits into the nucleus with a typical diameter of only 10 μm, and that nonetheless its information remains reliably accessible if required; information not required for a long time or the entire life of a specific cell has to be silenced in an effective way (Kepper et al. 2008). For example, genes required to be active during a given stage of development may have disastrous consequences if activated out of place. How are the profound differences in gene activities established and maintained in a large number of cell types to ensure the development and functioning of a complex multicellular organism (Cremer 1985; Cremer et al. 1988; Zuckerkandl 1997). To answer this question fully, in addition to nuclear biochemistry we need to understand how genomes are organized in the nuclei.

According to the role of a specific cell in the organism, different genes have to be activated and inactivated ("silenced") in a very precise way, from the "totipotent" fertilized oocyte to "pluripotent" embryonic stem, cells to adult stem cells, to "terminally differentiated" cells. An understanding of how the activation and silencing of specific genes occur would greatly facilitate e.g., the reprogramming of adult stem cells (i.e. cells found in the body of an adult organism). Towards this goal, a large amount of highly relevant molecular and structural information has been accumulated. Many valuable attempts have been made to integrate this wealth of knowledge into models of nuclear architecture. Presently, however, most models of this kind are 'qualitative'; this means that they describe certain general features of nuclear organisation, without trying to quantitate them. By necessity, also the conclusions drawn from such models are qualitative. For example, for a long time it has been noted that the concept of 'chromosome territories' and their spatial distribution in the cell nucleus is a key element to understand the formation of chromosome aberrations. Nevertheless, such ideas by themselves did not allow to make quantitative predictions about the formation of cancer related chromosome aberrations in a given cell type at a given dose of ionizing radiation. Such a lack of prediction is not restricted to radiation effects but to many other biologically and medically important features of functional nuclear architecture.

The general idea of cellular biophysics is to overcome this impasse (a) by gaining as much quantitative information about relevant features as possible; (b) by using such quantitative information to develop 'qualitative' models into quantitative ones, i.e. into models which allow quantitative, observable predictions of certain parameters; (c) by modifying the models in such a way that their predictive power is optimized. To be successful, this may require extensive efforts to develop the methods required to obtain the quantitative experimental information necessary for the predictive model calculations.

Concerning nuclear architecture and its biological and medical implications, until recently such a project was practically impossible to realize: On the experimental side, molecular information was missing; optical and labelling tools to perform quantitative analyses of nuclear structure on the level required were scarcely available; the computer hardware and the programmes to handle the huge amount of calculations were not existing.

In the following, some approaches to start such a programme towards quantitative nuclear architecture and modelling will be presented, based on the specific experience of the authors.

3.2 Experimental Evidence for Nuclear Genome Large Scale Architecture

Numerous studies have shown that the chromatin fibers of individual chromosomes in the cell nucleus are not distributed throughout the nucleus but there enveloping surface forms a volume which occupies only a relatively small part of the nucleus.

Figure 3.1 shows an example where all 24 chromosome types of the human genome were visualized by combinatorial Fluorescence-in situ Hybridization (FISH) (compare also Schroeck et al. 1996; Speicher et al. 1996).

The compartmentalization of the nucleus in several welldefined subregions such as nucleoli, nuclear bodies, chromosome territories (CTs), and their higher compartmentalization levels into subchromosomal domains as well as the spatial arrangements of these compartments may have a profound impact on functional processes inside the nucleus (for review, see Chevret et al. 2000; Dundr and Misteli 2001; Cremer and Cremer 2001; Parada and Misteli 2002; O'Brien et al. 2003; Kreth et al. 2004a; Cremer et al. 2000; Cremer et al. 2006; Cremer and Cremer 2006a,b; Dietzel et al. 1998; Albiez et al. 2006; Ferreira et al. 1997; Zirbel et al. 1993; Cremer et al. 1993; Visser et al. 2000; Solovei et al. 2002; Qumsiyeh 1999; Leitch 2000; Bridger 1999; Gonzalez-Melendi et al. 2000). For example, it has been shown that chromosome territories are compartmentalized into domains of early and later replicating chromatin (Visser et al. 1998; Zink et al. 1999; see also Tsukamoto et al. 2000): early replicating chromatin domains are found throughout the nucleus except for the utmost nuclear periphery and the perinucleolar space, whereas midreplicating chromatin domains form typical rims both along the nuclear periphery and around the nucleoli (Dimitrova and Berezney 2002). This specific arrangement of differently replicating chromatin may mirror the results of recent investigations, regarding the positioning of whole CTs inside the nuclear volume.

Figure 3.2 shows a 'qualitative' model of basic features nuclear architecture.

Chromosome painting experiments of single CTs and groups of CTs in different species suggest a relationship between the gene density of a chromosome and its radial positioning (distance to the nuclear center) in the nuclear volume. This was first shown

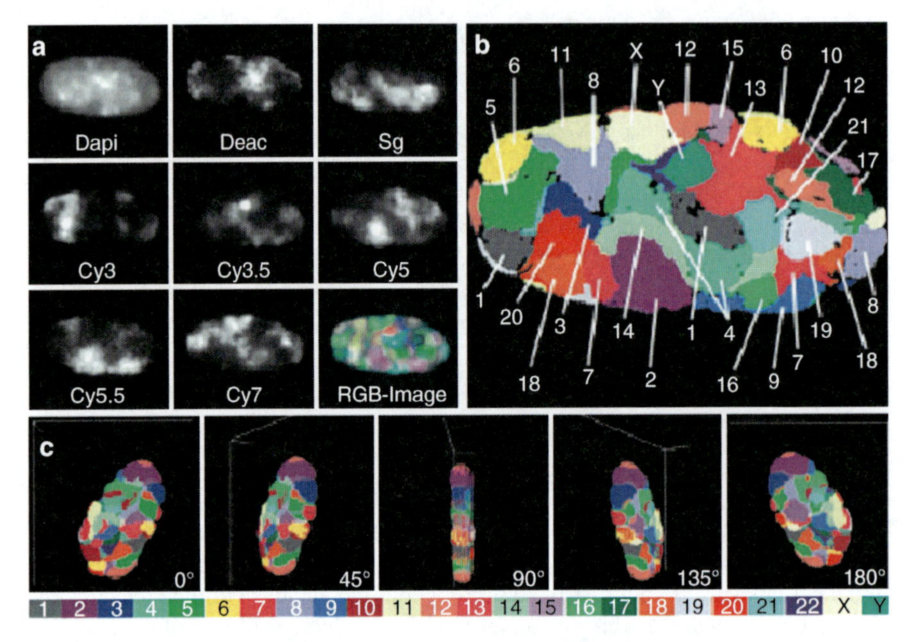

Fig. 3.1 24-Color 3D FISH representation ("Painting") and classification of Chromosome Territories (CTs) in a human G0 fibroblast nucleus. (**a**) A deconvoluted mid-plane nuclear section recorded by wide-field microscopy in eight channels: one channel for DAPI (DNA counterstain) and seven channels for the following fluorochromes: diethylaminocoumarin (Deac), Spectrum Green (SG), and the cyanine dyes Cy3, Cy3.5, Cy5, Cy5.5, and Cy7. Each channel represents the painting of a CT subset with the respective fluorochrome. Using a combinatorial scheme, pseudocolored images of the 24 differently labeled chromosome types (1–22, X, and Y) were produced by superposition of the seven channels (*bottom right*). (**b**) False color representation of all CTs visible in this mid-section after classification with a quantitative evaluation programme. (**c**) 3D reconstruction of the complete CT arrangement in the nucleus viewed from different angles (From Bolzer et al. 2005)

by Croft et al. in 1999 for the different positions of CTs Nos. 18 and 19 in human lymphocytes in a two-dimensional semiquantitative analysis: Both chromosomes are of similar DNA content, but the genepoor CT No. 18 was found at the nuclear periphery, whereas the gene-dense CT No. 19 was found in the nuclear interior.

3.3 Quantitative Microscopy of Nuclear Genome Architecture

A quantitative three-dimensional (3D) evaluation confirmed the positioning of the gene-dense CTs No. 19 toward the nuclear center and of the gene-poor CT No. 18 the nuclear periphery in morphologically preserved spherical nuclei of lymphocytes, which have an average diameter of 10 μm (Cremer et al. 2001). A gene density-correlated radial CT position for almost all chromosomes was described by Boyle et al. in 2001. Figure 3.2 shows an example for the quantitative measurement

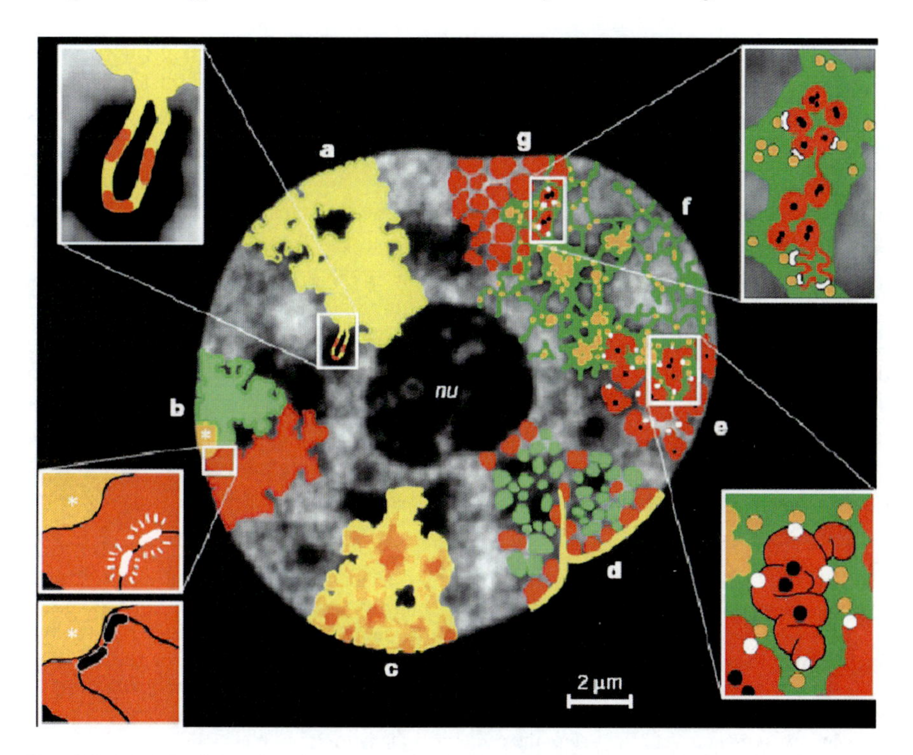

Fig. 3.2 Model of functional nuclear architecture. (**a**) CTs have complex folded surfaces. *Inset*: topological model of gene regulation23. A giant chromatin loop with several active genes (*red*) expands from the CT surface into the IC space. (**b**) CTs contain separate arm domains for the short (p) and long chromosome arms (q), and a centromeric domain (*asterisks*). *Inset*: topological model of gene regulation78,79. *Top*, actively transcribed genes (*white*) are located on a chromatin loop that is remote from centromeric heterochromatin. *Bottom*, recruitment of the same genes (*black*) to the centromeric heterochromatin leads to their silencing. (**c**) CTs have variable chromatin density (*dark brown*, high density; *light yellow*, low density). Loose chromatin expands into the IC, whereas the most dense chromatin is remote from the IC. (**d**) CT showing early-replicating chromatin domains (*green*) and mid-to-late-replicating chromatin domains (*red*). Each domain comprises ~1 Mb. Gene-poor chromatin (*red*), is preferentially located at the nuclear periphery and in close contact with the nuclear lamina (*yellow*), as well as with infoldings of the lamina and around the nucleolus (nu). Gene-rich chromatin (*green*) is located between the gene-poor compartments. (**e**) Higher-order chromatin structures built up from a hierarchy of chromatin fibres 88. *Inset*: this topological view of gene regulation27,68 indicates that active genes (*white dots*) are at the surface of convoluted chromatin fibres. Silenced genes (*black dots*) may be located towards the interior of the chromatin structure. (**f**) The CT–IC model predicts that the IC (*green*) contains complexes (*orange dots*) and larger non-chromatin domains (aggregations of *orange dots*) for transcription, splicing, DNA replication and repair. (**g**) CT with ~1-Mb chromatin domains (*red*) and IC (*green*) expanding between these domains. *Inset*: the topological relationships between the IC, and active and inactive genes 72. The finest branches of the IC end between ~100-kb chromatin domains. *Top*: active genes (*white dots*) are located at the surface of these domains, whereas silenced genes (*black dots*) are located in the interior. *Bottom*: alternatively, closed ~100-kb chromatin domains with silenced genes are transformed into an open configuration before transcriptional activation (From Cremer and Cremer 2001)

of radial positions obtained by confocal laser scanning fluorescence microscopy (Tanabe et al. 2002). Additionally, it could be shown that the distinct localization of the chromatin homologous to human chromosome No. 18 and of chromatin homologous to human chromosome No. 19, respectively, was maintained in lymphocytes during the evolution of higher primates, irrespective of major karyotype rearrangements that occurred in these phylogenetic lineages during their evolution, suggesting a functional significance for such an order (Tanabe et al. 2002).

However, the different positioning of a gene density related radial dependence of chromatin obviously does not apply for all human cell types. In nuclei of human diploid fibroblasts, the CTs of small CTs were found in the nuclear center irrespective of the gene density, while large chromosomes were positioned toward the nuclear periphery, arguing for a chromosome size rather than a gene density correlated radial arrangement (Cremer et al. 2003; Bolzer et al. 2005).

Figure 3.3 shows an example for the quantitative measurement of CT distributions in human cell nuclei obtained by first "painting" the chromosome territories to be

Fig. 3.3 (**a**) Three-dimensionally reconstructed CTs in human lymphoblastoid nuclei. Three-dimensional positioning of HSA18 (*red*) and HSA19 (*green*) CTs in a human lymphoblastoid cell nucleus with the partially reconstructed nuclear border (outside, *blue*; inside, *silver-gray*). HSA18-homologous CTs were always positioned close to the nuclear border. The relative locations of the CTs, however, varied from a close neighborhood to opposite positions. (**b**) Scheme of 3D evaluation of radial chromosome territory arrangements. A voxel (volume element)-based algorithm was applied. As a first step, the center and the border of the nucleus were determined by using the 3D data set of the DNA-counterstain fluorescence in the following way; first, the fluorescence intensity gravity center of the counterstain voxels after automatic thresholding was calculated. For the interactive segmentation of the nuclear border, a straight line was drawn from the gravity center toward each voxel considered, and the nuclear center was then determined as the geometrical center of the segmented voxels. In the second step, segmentation of CTs was performed in each 3D stack representing the color channels for the respective painted CTs. The segmented nuclear space was divided into 25 equidistant shells. For each voxel located in the nuclear interior, the relative distance r from the nuclear center was calculated as a fraction of r0.Ashell at a given r contains all nuclear voxels with a distance between $r - \Delta r/2$ and $r + \Delta r/2$. For each shell all voxels assigned to a given CT were identified and the fluorescence intensities derived from the respective emission spectrum were summed up. This procedure yielded the individual DNAshell contents for painted CTs as well as the overall DNAcontent reflected by the DNA counterstain. For better comparison of different nuclei, the sum of the voxel intensities measured in each nucleus was set to 100% for each fluorochrome. When this normalization is used, the average relative DNA content in nuclear shells as a function of the relative distance r from the 3D center represents the average radial distribution of the DNA representing the painted CTs or of the overall DNA in the entire set of evaluated nuclei. (**c**) Quantitative 3D evaluation of radial chromatin arrangements in primate cell nuclei with HSA18 and HSA19 homologues. Radial chromatin arrangements observed after painting with HSA18- and HSA19-homologous probes were evaluated in 25 radial concentric nuclear shells (compare (**b**)). The abscissa denotes the relative radius r of the nuclear shells, the ordinate denotes the normalized sum of the intensities in the voxels for a respective fluorochrome belonging to a given shell. For normalization, the area underlying the curve for each color (total relative DNAcontent) was set to 100. n_number of 3D evaluated nuclei. A highly significant difference was noted between the radial positioning of HSA18 (*red*) and HSA19 (*green*) homologous chromatin. HSA18-homologous chromosome material is consistently distributed closer to the nuclear border, whereas HSA19-homologous material is distributed toward the nuclear interior. Bars indicate standard deviations of the mean for each shell. Blue curves represent counterstained DNA (From Tanabe et al. 2002)

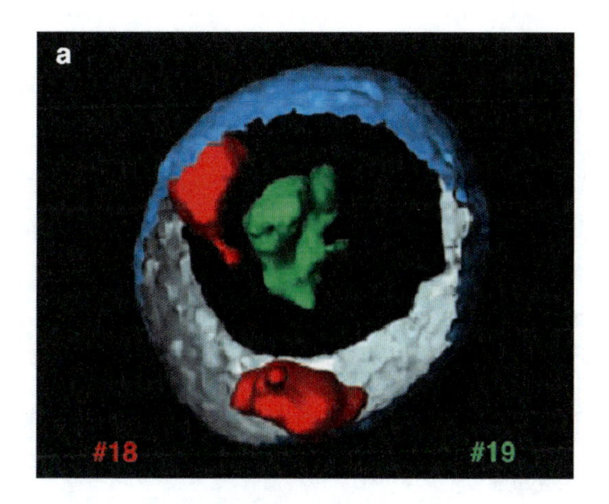
a
#18
#19

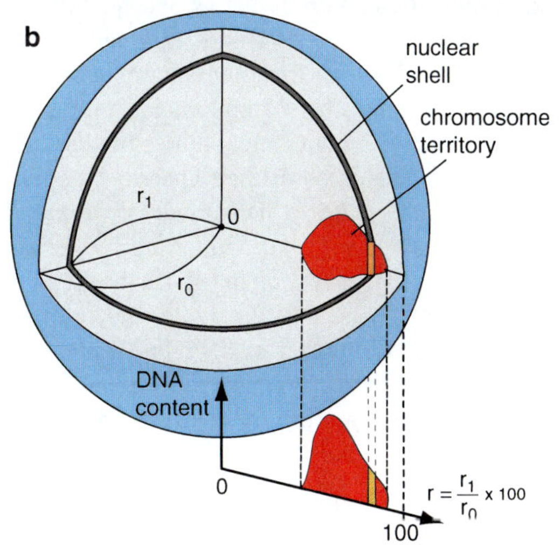
b
nuclear
shell
chromosome
territory
r_1
0
r_0
DNA
content
0
$r = \frac{r_1}{r_0} \times 100$
100

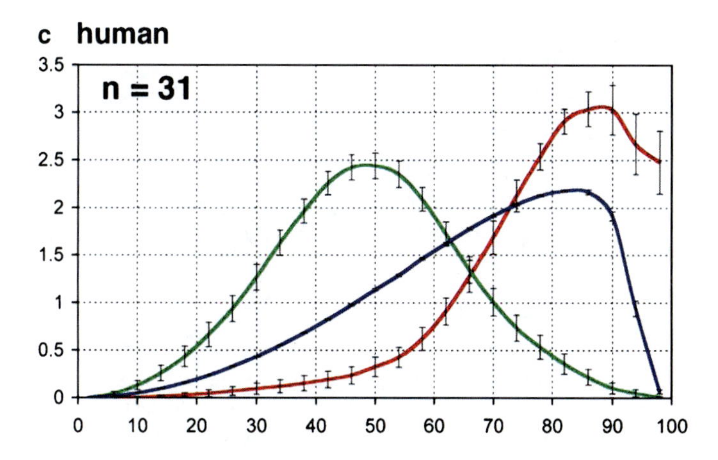
c human
n = 31
3.5
3
2.5
2
1.5
1
0.5
0
0 10 20 30 40 50 60 70 80 90 100

studied and then analyzing them by confocal laser scanning fluorescence microscopy (CLSM). Since their introduction, such quantitative measurements of chromatin distribution and other quantitative features have been amply used to convert qualitative information into quantitative ones. Only then, it became possible to compare nuclear structure with the quantitative predictions of computer models of nuclear architecture.

3.4 Quantitative Modeling of Nuclear Genome Large Scale Architecture

Every quantitative model of nuclear genome architecture (see e.g., Muenkel and Langowski 1998; Branco and Pombo 2007) has to be compatible with a number of well established experimental basic observations and quantitative results: (1) the existence of individual chromosome territories; (2) the fairly large variation of CT positions in the in the nuclei; (3) the distinct nonrandomness in the spatial distribution; (4) the existence of individual chromatin domains within the CTs. To explain these features of nuclear "macro" architecture, various models have been proposed. Here, we concentrate on the discussion of the "Spherical 1 Mbp Domain" (SCD) model presented by Kreth et al. in 2004a. This model approximates each chromosome of the diploid chromosome set (and additionally two nucleoli modeled as medium sized chromosomes) by a linear chain of 1 Mbp sized spherical domains with a diameter of 500 nm which are linked together by entropic spring potentials Different domains interact with each other by a slightly increasing repulsive potential (Fig. 3.4).

To maintain the territorial compaction of such a modeled chromosome chain, an additional weak enveloping potential barrier around the chain was necessary to

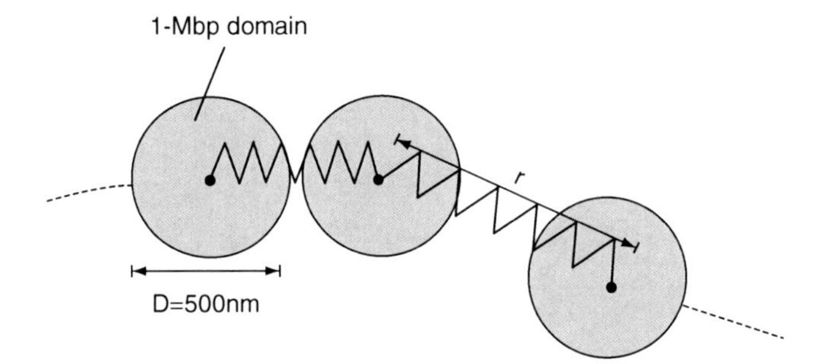

Fig. 3.4 Schematic drawing of the approximation of a chromosome by a linear chain of spherical 1 Mbp-sized domains, which are linked together by entropic spring potentials according to the spherical 1 Mbp chromatin domain (SCD) model (From Kreth et al. 2004a)

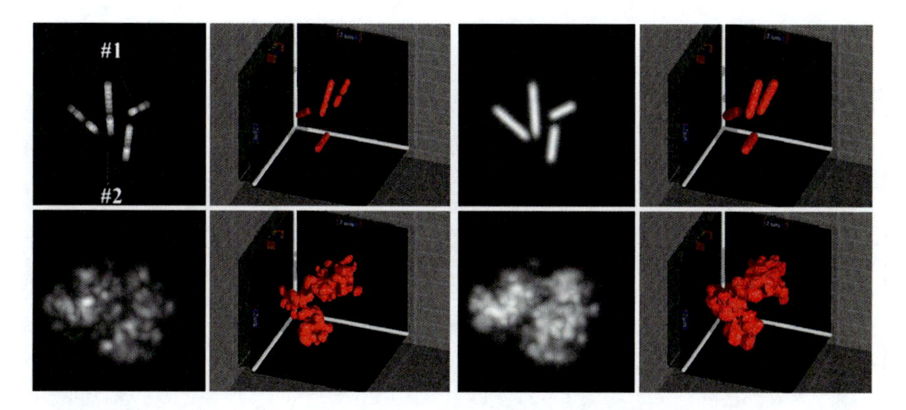

Fig. 3.5 Virtual microscopic images and reconstructions of modelled chromosomes #1, 2 according to the SCD model. Left panel shows a virtual G-banding of a mitose like (*upper row*) and a relaxed interphase like (*lower row*) configuration of both chromosomes. On the right side the same configuration is shown for a full painting. In each case on the left side an axial projection of the data stack is schown and on the right side a 3D reconstruction

prevent an intermingling of different chromosome chains in a semidilute polymer solution. Beginning from a mitotic like start configuration, a relaxation of the chromosome chains was performed until an equilibrium configuration was obtained. The transition from a start configuration to an energetic relaxed equilibrium state was performed by the Importance Sampling Monte Carlo procedure (Metropolis et al. 1953). The start configurations were made in such a way that for lymphocyte nuclei a gene density correlated arrangement of chromosome territories (CTs) in the nuclear volume was obtained (that means gene rich CTs are localized more in the interior and gene poor CTs more in the exterior of the nuclear volume) and for fibroblast nuclei a size (DNA content) dependent arrangement of CTs. In both cases the distribution of CTs in the nuclear volume has a probabilistic and not a deterministic character (compare Kreth et al. 2004a). Based on the 850 ideogram banding pattern each 1 Mbp domain was then assigned a label, which identifies the domain as an R-band/G-band/C-band (centromer) or a nucleolus domain (compare Fig. 3.5).

As a first example for the application of the SCD model , the radial distributions (distance from the nuclear interior to the nuclear border) of single CTs #12, 18, 19, 20 in spherical lymphocytes were computed (Cremer et al. 2001; Weierich et al. 2003). Chromosome 12 (142 Mbp [International Human Genome Sequencing Consortium 2004]) and chromosome 20 (66 Mbp) represent chromosomes with intermediate gene densities, while chromosome 18 (86 Mbp) represents a gene poor chromosome and chromosome 19 (72 Mbp) the most gene dense gene human chromosome.

According to the "Spherical 1 Mbp Chromatin Domain (SCD)" model the simulated chromosome chains consisting on a certain number of spherical 1 Mbp-domains (according to the DNA content of a chromosome) which were arranged at the beginning (start configuration) in mitotic like "start cylinders". The model calculations were based on three different assumptions about the initial distribution

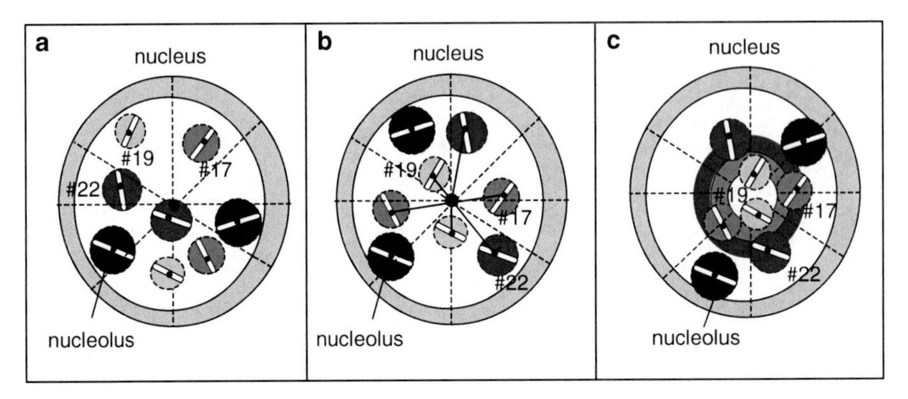

Fig. 3.6 Schematic drawing of the localization of the initial CT spheres in the nuclear volume for the three simulated cases. In the statistical simulation case (**a**), the initial CT spheres were put in the nucleus in a random order without further assumptions. In the probabilistic simulation case (**b**), the initial CT spheres were put in the nucleus in the order of their gene densities, and the distances of the CT spheres to the nuclear center were weighted with a probability density function according to their gene densities. In the deterministic simulation case (**c**), the initial spheres were located on discrete shells in the order of their gene densities. Starting with the initial spheres of CT No. 19 on the first shell in the interior, the next two CTs, No. 17 and No. 22, follow in the upper shells and so on. A constraint that has to be fulfilled in all three cases is that overlapping of the initial CT volumes is not allowed (From Kreth et al. 2004a)

of these "start cylinders": ***statistical***, ***deterministic*** and ***probabilistic*** distribution (see Fig. 3.6). In addition, two nucleoli were inserted in all three cases, simulated as additional CTs with a DNA content of 80 Mbp. The midpoints of the nucleoli in the start configuration were considered to maintain a minimal distance of 1.75 μm to the nuclear envelope and a minimal distance of 3.75 μm from each other. The "start cylinders" were located first in so called "initial" CT spheres. In the case of the ***statistical*** distribution of the CTs (see Fig. 3.6a) the "initial" spheres were positioned randomly in the nuclear volume with the condition that overlapping with still existing "initial" spheres was forbidden. As a consequence, in case of an overlap of a randomly chosen position of a given "initial" CT sphere with another CT sphere, this position was discarded, and a new random position was chosen. This procedure was repeated, until a non-overlap position was obtained. To let the algorithm converge (i.e. all "initial" CT spheres find an non-overlap position), the volumes of the "initial" spheres had to be reduced by a common factor v.

In the case of the ***deterministic*** and ***probabilistic*** distribution, a gene density correlated distribution of the "initial" CT spheres in the nuclear volume was performed. To create the ***deterministic*** start distribution (Fig. 3.6c) after the incorporation of the two nucleoli, the "initial" spheres of the homologous CTs were located on discrete shells in the nuclear volume in the order of their gene densities as following: #19, 17, 22, 11, 1, 15, 14, 12, 20, 6, 10, 3, 7, 2, 16, 9, 5, 21, 8, 4, 13, 18, X, Y. The simulation (for details see Kreth et al. 2004a) was started with the "initial" spheres of the CTs with the maximum gene density (CTs #19); then the CTs

#17 spheres with the second highest gene density were located with an appropriate distance from the first shell and so on. In this deterministic start distribution, all probabilistic constraints were eliminated, except that on a given shell surface, an "initial" CT sphere was allowed any radial position not resulting in an overlap.

For the ***probabilistic*** case (Fig. 3.6b), after the incorporation of the two nucleoli, the CTs were put into the nuclear volume in the same order according to gene density as realized for the deterministic case. Here, however, in contrast to the deterministic case, the "initial" CT spheres were not located on discrete shells but the distance from the center of the "initial" spheres to the nuclear center was weighted with an exponential probability density function which depends on the gene density of a given chromosome. A non overlapping position of the CT in the nucleus was chosen. Then the distance d to the nuclear center was determined for this special position, and a probability vaule P(d) was calculated with the gene density of CT A (e.g., #22). Then the calculated $P(d)$ value was compared with a random number between 0 and 1. If the random number turned out to be equal or larger than the calculated $P(d)$ value, then the position of the "initial" (CT A) sphere was accepted. If the random number turned out to be smaller than the calculated $P(d)$ value, then again a new randomly chosen position for CT A was tested for non overlap; the d value of the new non overlapping position was again inserted in Eq. (3.8) and tested as described above. The procedure was continued until a non-overlap position was obtained with a random number equal/larger than the $P(d)$.

For the gene density values applied for each CT, sequence data derived ratios were used of observed versus expected gene-based ETS markers per chromosome (Deloukas et al. 1998).

After the start configuration with the "initial" spheres of the diploid human chromosome set (22, X, Y) and the two nucleoli had been created as described above, the "start cylinders" were placed inside these spheres. In the next step, for the relaxation process of the "start cylinders" into an equilibrium state, the "initial" spheres were discarded and played no further role in the relaxation process. For all three cases, 50 nuclei each were calculated. For comparison of the experimentally observed and simulated radial arrangements of the reconstructed CTs #12,18,19, and 20 the simulated nuclear configurations were virtually labeled using a "virtual microscopy" approach. Figure 3.7 visualizes 3D reconstructions of painted CTs #18 and 19 in a nucleus of a human lymphocyte (Fig. 3.7d) as well as for the three simulated model assumptions (Fig. 3.7a–c).

The quantitative 3D evaluation of the nuclear positioning of the (virtually) painted territories was made by the assessment of the 3D relative radial distribution of each voxel assigned to the respective territory. Figure 3.8 shows the voxel distributions for the respective CTs plotted against the relative radius in lymphocyte nuclei for the probabilistic case.

In the statistical simulation case, both CTs #18 and 19 had a very similar (peripheral distribution) which was in clear contrast with the experimental results (compare Fig. 3.7b). In the deterministic simulation case, the correlation of simulation and experiment was much more pronounced. In the probabilistic simulation case, the evaluated more interior arrangement (in the nuclear volume) of the CTs No. 19, and

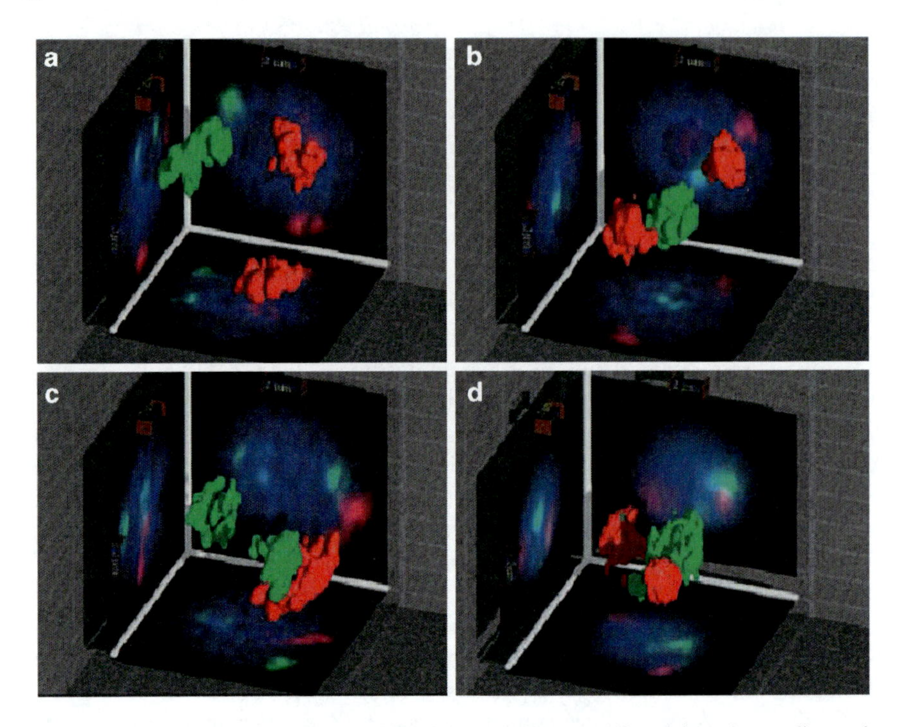

Fig. 3.7 Visualization of reconstructed CTs of simulated human cell nuclei (**a**–**c**) according to the SCD model and of an experimental human lymphocyte cell nucleus with FISH-painted CTs. (**d**) The simulated virtual microscopy data stacks are reconstructions from the three simulation cases of the relaxed configurations: statistical simulation case (**a**), probabilistic simulation case (**b**), and deterministic simulation case (**c**). In all cases, CTs No. 18 were visualized in red and CTs No. 19 in green. The visualization tool was kindly provided by Dr. R. Heintzmann, University Jena,Germany (From Kreth et al. 2004a)

the more peripheral arrangement of the CTs No. 18, fitted the experimental data best. These examples (Kreth et al. 2004a) indicate that already relatively basic quantitative models allow to predict nuclear genome macrostructure (here the radial distribution of specific CTs) in a promising way.

3.5 Application of the SCD Computer Model to Predict Cell Type Specific Radiation-Induced Chromosomal Aberrations

The non random higher order spatial arrangement of chromosome territories (CTs) and specific genomic regions in eukaryotic cells significantly contributes to their likelihood of undergoing chromosomal aberrations once chromosome breaks have occurred (Kozubek et al. 1999; Cornforth et al. 2002; Roix et al. 2003; Arsuaga et al. 2004;

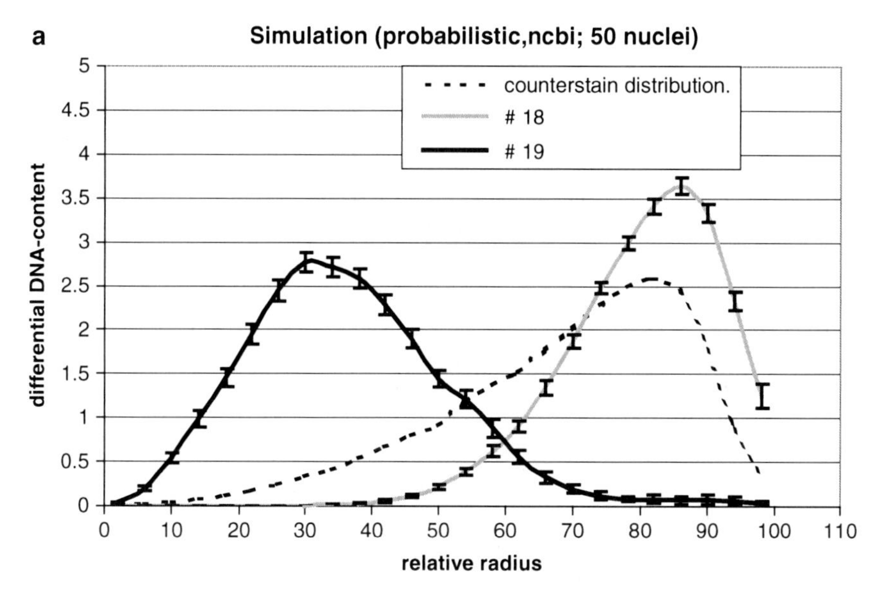

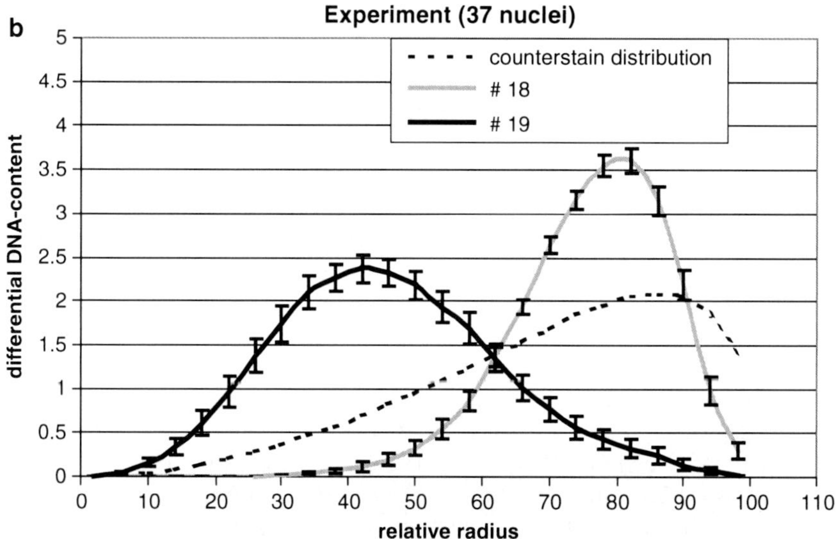

Fig. 3.8 Radial distribution curves of experimental (compare Cremer et al. 2001; Weierich et al. 2003) and quantitative computer simulations applying a 3D mapping algorithm based on the SCD model. The radial arrangements were evaluated human CTs No. 18 and No. 19. The counterstain distribution results from the mapping of all chromosomes. (**a**) Simulation results; (**b**) Experimental results for the radial distribution of #18 and 19. The relative radius determines the relative position of a shell with respect to the nuclear border. For example, a shell at the relative radius 0 is located at the nuclear center, whereas the shell 98 is positioned at the nuclear periphery. Error bars represent the standard deviations of the mean. The mean value for each relative radius was obtained by the average of the single distribution curves for each nucleus. (From Kreth et al. 2004a)

Parada et al. 2004). In lymphocytes, CTs are arranged in a radial manner which is evolutionarily conserved in lymphocytes (Tanabe et al. 2002) and has also been reported in fibroblasts. However, while in lymphocyte type cell nuclei the gene density seems to be the underlying factor arranging gene dense chromosomes in the interior and gene poor chromosomes in the periphery of the nuclear volume (e.g., Boyle et al. 2001; Cremer et al. 2001) , in fibroblast nuclei a high correlation with the size was reported (Sun et al. 2000; Bolzer et al. 2005). Radial positioning is also not limited to entire chromosomes, but has also been reported for single genes (Roix et al. 2003). For all these experimentally observed radial patterns, the probabilistic nature is a common further feature. That means, although average positions of genes and chromosomes, relative to the center of the nucleus or relative to each other, can be measured in absolute terms, the position of a single gene region or chromosome varies greatly among cells in a population.

To evaluate specific aspects of radiation induced aberrations, quantitative models based on nuclear architecture have been applied (Sachs et al. 1997; Sachs et al. 2000; Arsuaga et al. 2004). In these models regard, a geometric representation of the chromosomes in a given nuclear volume was used (Holley et al. 2002; Friedland et al. 2008;. Ballarini et al. 2002). For example, to attend on the lower order chromatin level break point formations, Holley et al. (2002) described the CT structure by a polymer, in which the 30 nm fiber was allowed to form a three-dimensional random walk within a spherical chromosome volume. To construct irregularly shaped CT volumes, Ballarini et al. (2002) presented a model where the nuclear volume was first divided in a 3D grid of 27,000 cubic elements (boxes) and CTs were constructed then by subsequent occupation of closest neighboring boxes. In both model approaches the volumes of CTs were proportional to their DNA contents; a specific arrangement of gene dense or gene poor chromosomes in the nuclear volume or a size dependent arrangement however was not regarded.

In an effort to quantitatively estimate the chromosomal aberrations under the assumption of a non random chromosomal arrangement, we implemented the Spherical 1-Mbp chromatin domain (SCD) model, assuming a preferential distribution CTs corresponding to their gene densities in interphase lymphocyte nuclei (compare Fig. 3.9) (Kreth et al. 2004a). Based on such calculated nuclear configurations, pairwise aberration frequencies were determined for Low LET (Linear Energy Transfer) irradiation (Kreth et al. 2004b). Clear differences to a calculated statistical distribution of CTs in the nuclear volume were observed. In comparison with an experimental study of Cornforth et al. (2002), the influence of proximity effects on interchange yields was confirmed. Here we give an example for the application of the SCD computer model described above to predict cell type specific radiation-induced chromosomal aberrations.

On the assumption of a random distribution of double strand breaks (DSBs) within the DNA for a Low LET radiation, the probability of a break occurring within a certain 1 Mbp domain was modeled using Poisson distribution mathematics. This assumes that, although an ionizing radiation track may produce multiple DSBs, these are distributed randomly throughout the genome. Under the assumption that the number of DSBs induced within a nucleus increases linearly with

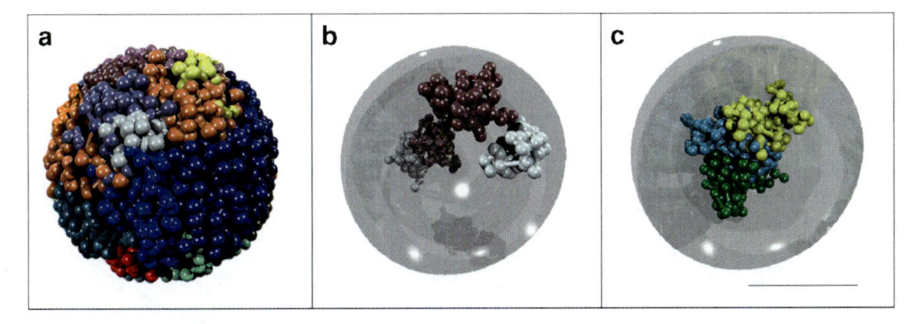

Fig. 3.9 Visualization of a modeled lymphocyte nucleus according to the SCD model. (**a**) Visualization of the complete diploid chromosome set of the human genome after 400,000 Monte Carlo steps. Each homologous chromosome pair is labeled in a separate color. (**b**) The two gene poor CTs #18 and 21 are visualized only. The peripheral organization is clearly visible. (**c**) The three gene rich CTs #17, 19, 22 are visualized; here the internal positioning near the nuclear center is preferred according to the experimental observations. The bar denotes 5 μm (From Kreth et al. 2007)

dose and is proportional to the DNA content of the cell, the probability p_n of an individual modeled 1 Mbp domain containing *n* DSBs was calculated from an adaptation of the equation of Poisson distribution (Johnston et al. 1997).

The DSBs within the 1 Mbp domains were placed randomly. To determine an exchange (inter-/intra-change) between two DSBs in two 1 Mbp domains, only those 1 Mbp domains containing DSBs were regarded which revealed a distance (the distance between the mid points of both 1 Mbp domains) smaller or equal to a certain proximity value. Besides the proximity value of 500 nm (Kreth et al. 2004b), also a proximity value of 250 nm (which means that the 1 Mbp domains must overlap) and also proximity values of 750 and 1,000 nm were tested. An exchange event in dependence of the distance *d* between the two DSBs in the two domains was counted according to the normalized probability function $p_d = [r/d]^a$. Here, *r* denotes the radius of a 1 Mbp domain which determines the maximal distance by which an exchange takes place in every case. For the exponent *a* different values were tested and compared with experimentally obtained dose response curves (Fig. 3.10). When for a certain DSB an exchange was not counted, other domains in a certain distance (proximity value, see above) containing DSBs were tested. When this procedure failed, the DSB was considered as repaired. Exchanges between domains of the same chromosome were counted as intrachanges and were separated from interchanges (For details see Kreth et al. 2007).

These application examples show that the prediction of radiation induced chromosome aberrations in human lymphycyte can be highly improved on the basis of computer models taking into regard quantitative structural information. With a reasonable parameter set, the Spherical 1 Mbp Chromatin Domain (SCD) model allowed an excellent agreement (with respect to the amount and the behavior) of calculated dicentrics/translocation frequencies with experimental dose response curves in a

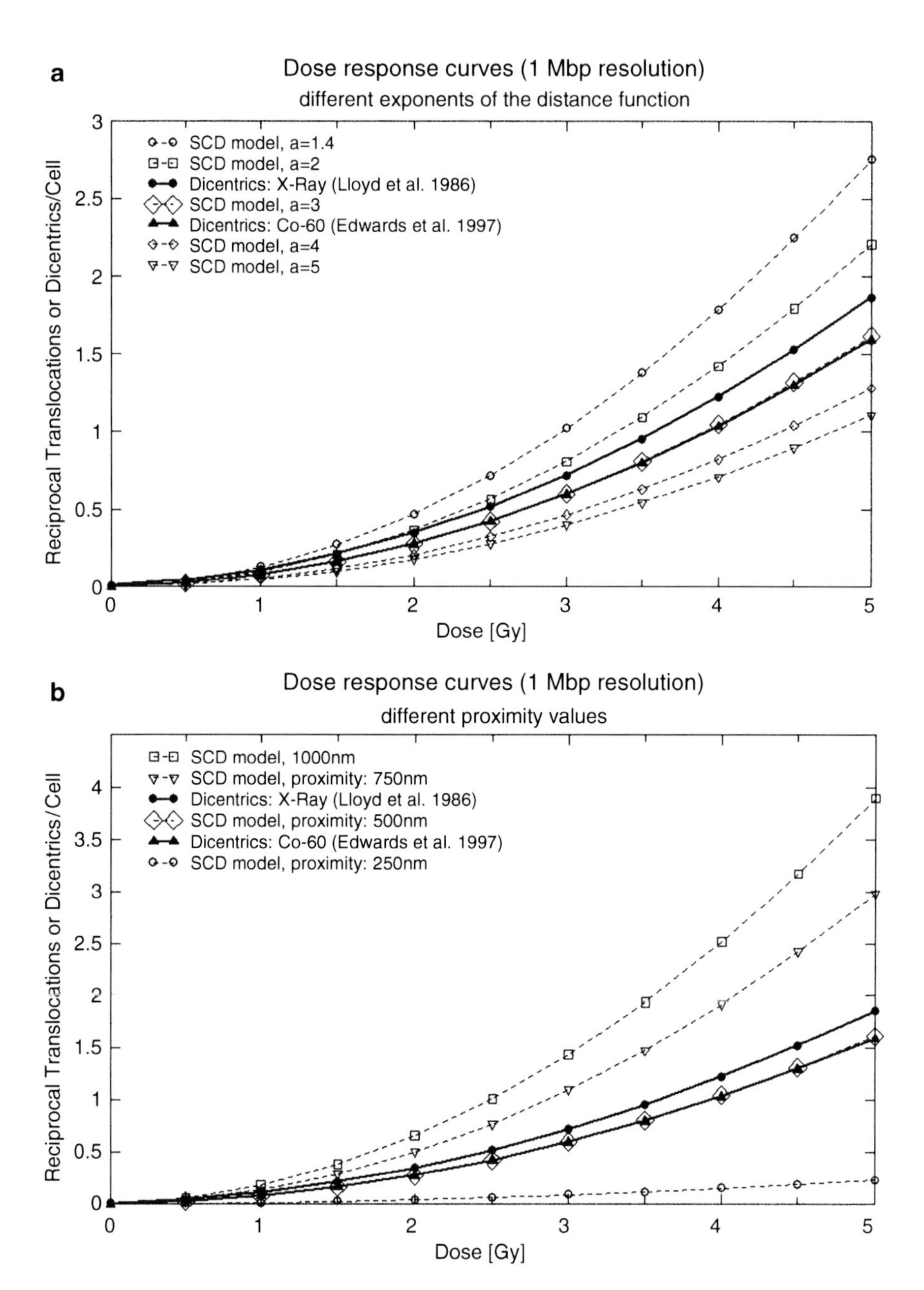

Fig. 3.10 Calculated dose response curves (with the SCD model) for dicentrics/translocations per cell were compared with measurements of Lloyd et al. 1986 and Edwards et al. 1997. To represent the experimental measurements, their linear-quadratic expressions ($0.005 + 0.036D + 0.067D^2$) and ($0.018D + 0.060D^2$) were used. Also for the representation of the calculated dose response curves, only the linear-quadratic fitting curves are shown. (**a**) For these calculations the proximity value was set to 500 nm and the exponent of the probability distance function was varied (a = 1.4, 2, 3, 4, 5). (**b**) For these calculations the exponent a = 3 was fixed and the proximity value was varied (250, 500, 750, 1,000 nm). The best agreement was found for a = 3.0, and proximity value = 500 nm. The respective linear-quadratic expression was $0.015D + 0.061D^2$ (From Kreth et al. 2007)

Table 3.1 Interchange frequencies of specific chromosome pairs calculated on the basis of the SCD computer model were compared with the respective experimentally observed frequencies of irradiated peripheral lymphocyte blood cells (Data from Arsuaga et al. 2004). The calculated frequencies are shown for a gene density correlated and a statistical arrangement of CTs in spherical nuclei. For both simulation cases 1,000,000 cell configurations were taken (From Kreth et al. 2007)

	Calculated absolute interchange frequencies in percent; 1,000,000 cell configurations; 3Gy Low LET		
Specific chromosome pair	Gene density correlated arrangement of CTs (%)	Statistical arrangement of CTs (%)	Human peripheral blood lymphocyte cells (Arsuaga et al. 2004), 3,585 cells; 2–5Gy Low LET (%)
f(4;18)	0.25	0.35	0.31
f(19;18)	0.051	0.19	0.056
f(19;17)	0.94	0.19	0.11
f(17;18)	0.095	0.23	0.14
f(9;22)	0.27	0.23	0.28
f(1;22)	0.46	0.37	0.75
f(1;21)	0.41	0.34	0.25

range between 0.5–5 Gy. Calculating a high number of nuclear structure configurations (one million), it was possible to predict absolute interchange frequencies for radiation doses in the 2–5 Gy range.

For radiation doses in the 2–5 Gy range. Although such doses are highly damaging in whole body exposures, even much higher doses are used in tumor treatment in partial body irradiation. Table 3.1 shows that in the important case of 9;22 translocations correlated to chronic myeloic leukemia, the SCD model results in very good predictions. In other cases, the correlation between prediction and observation was not as obvious, indicating that additional features, e.g., different compaction levels of CTs, differential repair rates, or selections against certain aberrations have to be regarded.

3.6 Extension of the SCD Model of Large Scale Nuclear Genome Architecture to Simulate Interactions with Other Nuclear Bodies

Besides the chromatin containing compartments, the nucleus is compartmentalized into specific protein-rich domains such as nucleoli, Cajal bodies and promyelocytic leukaemia bodies and nuclear speckles or splicing factor compartments, (reviewed by Lamond and Earnshaw 1998). The term nuclear speckle is used for aggregations of pre-mRNA splicing factors in mammalian cell nuclei that correspond to interchromatin granule clusters (IGCs) at the electron microscopic level and that is surrounded by perichromatin fibrils, which represent nascent transcripts of individual genes (Xing et al. 1995).

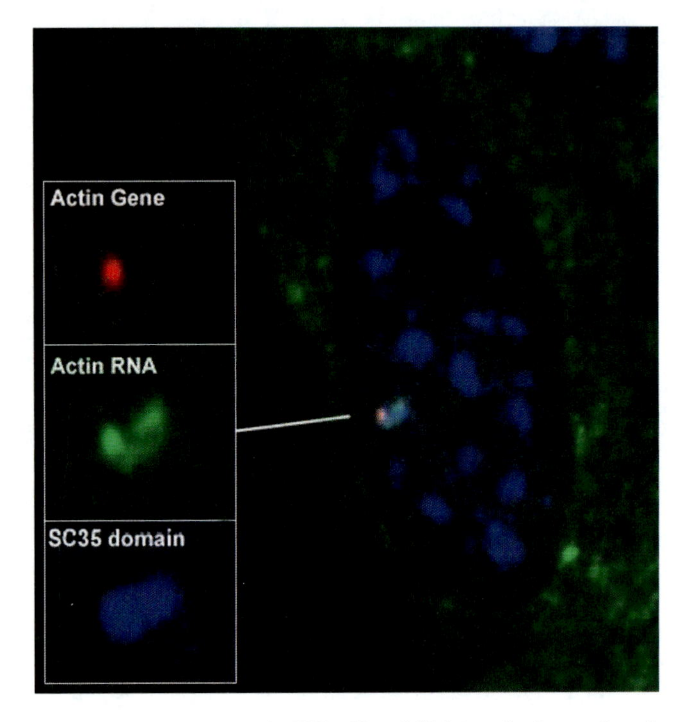

Fig. 3.11 *Red*: Actin-gene; *green*: Actin-RNA; *blue*: SC35-domain (Image from http://redbone.umassmed.edu/SC35.html)

Pre-mRNA splicing factors are not homogenously distributed over the whole nuclear plasma but are enriched in 20–50 nuclear locations with diameters between 0.5 and 3 μm. At the remaining space of the nuclear plasma these factors are present but only at a very low concentration. Locations with a very high concentration are called speckles (Fig. 3.11).

One of these splicing factors is the spliceosome assembly factor SC35. Speckles containing this factor are called also SC35 domains or splicing factor compartments. Interpreting the formation of SC35-domains two model presentations are discussed:

Model 1: mRNA metabolic factors, including SC-35 accumulate on transcripts of a single highly active gene or genomic clusters of genes. This model requires no structural organization of genes relative to the large concentrations of splicing factors. Such a model presentation is closely related to the thermodynamic switch model of nuclear architecture proposed recently by Nicodemi and Prisco 2009. This model supports the view that a variety of intra and inter-chromosome interactions can be traced back to similar mechanisms. Looping and compaction, remote sequence interactions, and territorial-segregated configurations correspond to thermodynamic states selected by appropriate values of concentrations/affinity of soluble mediators (e.g., mRNA metabolic factors) and by number and location of their attachment sites along the chromatin fiber. At regions of multiple attachment sites a larger number of factors will accumulated which might result then in aggregations of SC-35 domains.

Model 2: Multiple genes cluster at the periphery of a single large accumulation of mRNA metabolic factors. R-band DNA, which is gene rich, is more intimately associated with these SC-35 domains than gene poor G-band DNA. This view is experimentally supported by studies of FISH labelled R-bands, SC-35 domains and a subset of genes (Shopland et al. 2003). The studies revealed that each chromosome territory associates with three or four SC-35 domains, indicating specialized regions at the chromosome territory periphery. These contain domain-associated genes that can even come from different chromosome arms. An SC-35 domain can also associate with genes from different chromosomes. Because domain choice for individual genes is often random, the relative positions of their respective chromosomes also may be highly variable. The clustering of multiple specific genes and extensions of R-band DNA around the periphery of SC-35 domains reflects an organization around a structure, rather than of factors simply collected on the transcripts of a highly expressed gene.

To translate these experimental findings in a quantitative model representation in the present contribution the 1 Mbp Spherical Chromatin Domain (SCD) model was extended to regard attractive interactions of gene rich R-band domains with pre existing spherical SC35 domains according to Model 2. This approach (presented here for the first time)allows it for the first time to evaluate the influences from protein – chromatin interactions on the higher order nuclear architecture.

3.6.1 SCD model and banding pattern

To model the interaction between chromatin (the 1 Mbp SCD R/G band domains of chromosome territories), additional interactions between the introduced SC35-domains and the respective R-band/G-band domains were introduced. For the realization of combined attractive and repulsive interactions the Lennard Jones potential is commonly used. This potential consist on the attractive part $-\left(\frac{\sigma}{x}\right)^6$ and the repulsive part $\left(\frac{\sigma}{x}\right)^{12}$. The following interactions were implemented:

SC35-domain and R-band domain (attractive and repulsive):

$$V_1(x) = 4\varepsilon_1 \cdot \left(\left(\frac{\sigma_1}{x}\right)^{12} - \left(\frac{\sigma_1}{x}\right)^6 \right)$$

SC35-domain and G-band domain (repulsive):

$$V_2(x) = 4\varepsilon_2 \cdot \left(\left(\frac{\sigma_2}{x}\right)^{12} - \left(\frac{\sigma_2}{x}\right)^6 \right) + \varepsilon_2 \text{ if } x \leq r_{\min,2}$$

$$V_2(x) = 0 \text{ if } x > r_{\min.2}$$

SC35-domain and SC35-domain (repulsive):

$$V_3(x) = 4\varepsilon_3 \cdot \left(\left(\frac{\sigma_3}{x} \right)^{12} - \left(\frac{\sigma_3}{x} \right)^{6} \right) + \varepsilon_3 \text{ if } x \le r_{\text{min},3}$$

$$V_3(x) = 0 \text{ if } x > r_{\text{min},3}$$

The potential depth ε describes the strength of the potential between the respective interacting domains at distance x, σ the null position and r_{min} the minimum of the potential.

For the model calculation a mean number of 30 spherical SC35 domains with a mean diameter of 1.5 μm were introduced. The values for the potential strength ε_2 between two speckles and for the potential strength ε_3 between G-band domains and speckles were chosen to correspond to 1.5 kT which is identical with the interaction potential strength ε_0 between two 1 Mbp domains. This value ensured that a complete penetration of different domains during the relaxation process was prevented (Fig. 3.12).

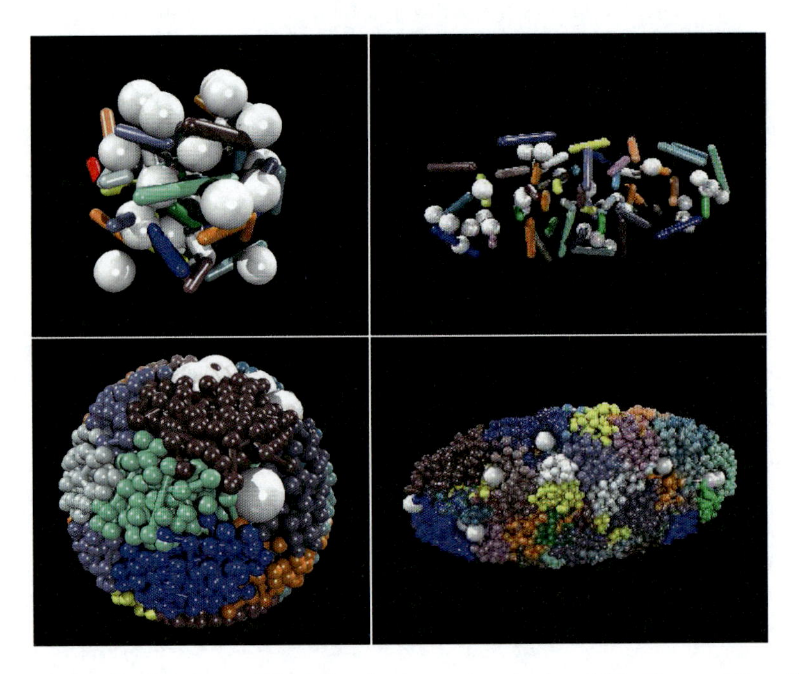

Fig. 3.12 Mitotic like start configuration (*upper row*) and relaxed interphase configurations (*lower row*) of a spherical shaped lymphocyte nucleus (*left side*; 10 μm in diameter) and an ellipsoidal shaped fibroblast nucleus (*right side*; 20 μm in diameter) together with 30 SC35 domains are shown (visualized in *white*). A SC35 domain has a diameter of 1.5 μm

3.6.2 Calibration

The value of the parameter ε_1 of the interaction potential between R-band domains and SC 35 domains was varied between $\varepsilon_1 = 0.0$ and $\varepsilon_1 = 12.0$ kT, where as for the range $\varepsilon_1 = 0.0$–2.0 kT a smaller step width was chosen and for the range ε_1=2.0–12.0 kT a larger step width. For each chosen value of ε_1 an ensemble of four relaxations was calculated. To obtain an optimal setting for ε_1 in the model which satisfy best the experimental observations two criteria were chosen:

Criteria I: all contacts between G-band domains with SC35 domains (Gbandcounts) and all contacts between R-band domains with SC35 domains (Rbandcounts) were summed up and the ratio **G2RContacts** was determined as the quotient between Gbandcounts and Rbandcounts. This ratio was experimentally observed by Shopland et al. (2003) to a value of 1/8.

Criteria II: The mean number of contacted SC35 domains per CT was saved in the variable **contPerTer**. Mean values for this variable between 3 and 4 were experimentally observed by Shopland et al. (2003).

Figure 3.13 showed the variation of the potential strength ε_1 in units of kT. In the upper graph of the figure the variable **contPerTer** was plotted as a function of ε_1. Two horizontal lines at y = 3 and y = 4 in green visualized the allowed region according to criteria II. The projection of the intersection points of the horizontal lines with the **contPerTer** curve on the abscissa revealed the allowed region for ε_1 (blue vertical lines).

This resulted in:

$$\varepsilon_{1,\min} = 1.35377 \pm 0.03799\text{kT}$$

$$\varepsilon_{1,\max} = 4.06977 \pm 0.09477\text{kT}$$

In the second lower graph of Fig. 3.13 the ratio **G2RContacts** was plotted as a function of ε_1. A horizontal line with y = 1/8 visualized the experimentally observed value according to criteria I. At the middle part of the curve an exponential fit was adapted and the intersection point of the line y = 1/8 with this fit resulted in the optimal value for ε_1 according to criteria I.

The upper and lower bounds of the error bars were also fitted and the intersection points of these fits with the line y = 1/8 were calculated; the projection on the abscissa resulted in an upper and lower error estimation of $\varepsilon_{1,\text{opt}}$.

$$\varepsilon_{1,opt} = 1.90961^{+0.4248}_{-0.4616}\, kT$$

The comparison revealed that $\varepsilon_{1,\min} < \varepsilon_{1,\text{opt}} < \varepsilon_{1,\max}$. This means that $\varepsilon_{1,\text{opt}}$ was located in the allowed region and both criteria were compatible.

A recalculation of $\varepsilon_{1,\text{opt}}$ with T = 310 K in energy units revealed :

$$\varepsilon_{1,opt} = 1.90961^{+0.4248}_{-0.4616}\, kT = 8.17^{+1.82}_{-1.97} \cdot 10^{-21} J = 0.051^{+0.011}_{-0.012}\, eV = 1.17587^{+0.261}_{-0.284} \frac{kcal}{mol}$$

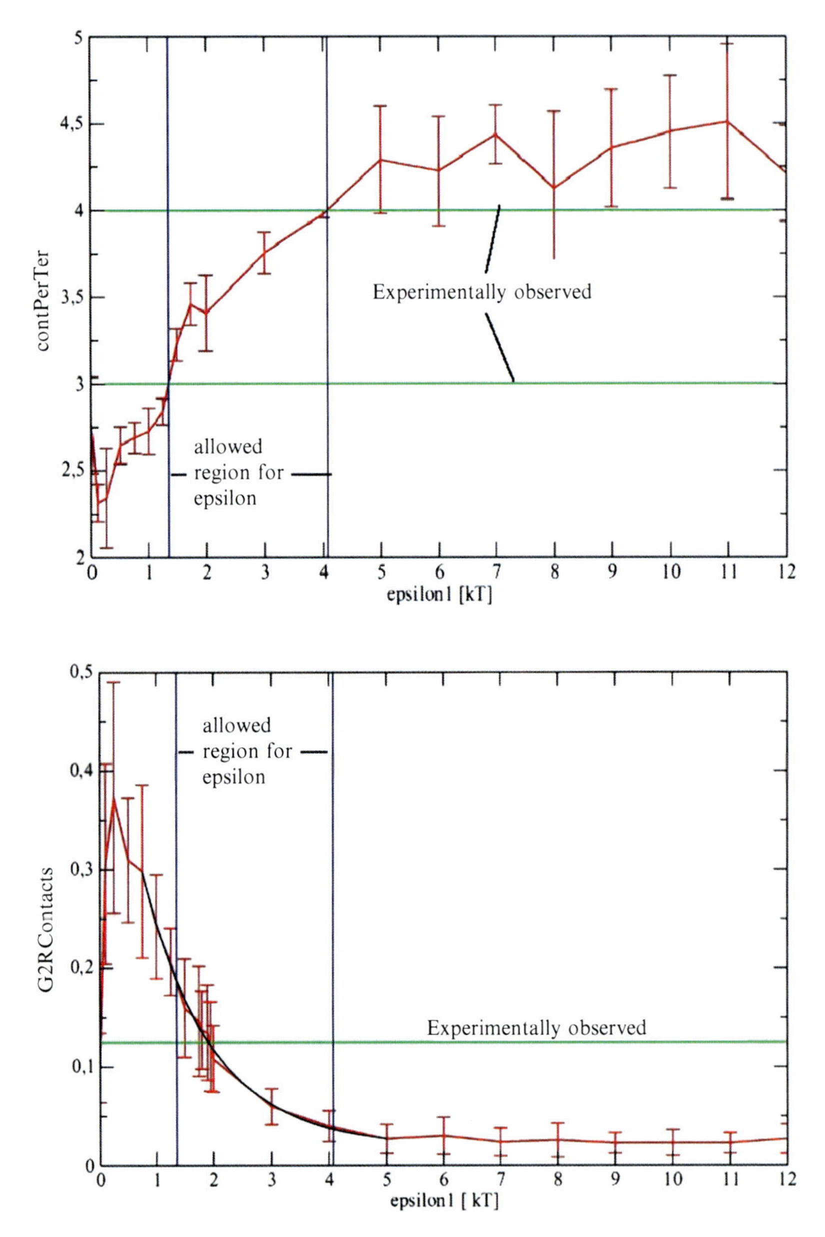

Fig. 3.13 Variation of the parameter ε_1 of the interaction potential between R-band domains and SC35-domains in units of kT. The upper graph visualizes the mean number of contacted SC35 domains per CT contPerTer. The green lines show the experimentally observed region of contacts (Shopland et al. 2003) according to criteria II. The blue lines result from the intersection points of the green lines with the calculated variation curve and visualize the allowed region for ε_1. The lower graph shows the relation G2RContacts of G band contacts divided by R band contacts. The green line visualizes the experimentally observed value (Shopland et al. 2003) and the blue lines visualize the allowed region from the upper graph

Hydrogen bonds in solution have a binding energy of:

$$\varepsilon_H \sim 1\frac{kcal}{mol}$$

Comparing both values one can conclude in a very rough first estimation that the interaction strength between R-band domains and SC35 domains might be in the order of only 1.18 hydrogen bonds, despite of the quite large contact surface which would allow a much larger number.

In a second calculation the size of the SC35 domains was reduced from 1.5 to 1.0 μm. In this case the value $\varepsilon_{1,opt}$ was reduced slightly to $\varepsilon_{1,opt}$ = 1.56139 kT. However it could be shown that this reduced size for a S35 domain is not longer compatible with the experimentally observed criteria I, II (calculations were not shown here).

3.6.3 Comparison with Experimental Observations

Taking into account the calculated optimal value $\varepsilon_{1,opt} = 1.90961^{+0.4248}_{-0.4616}kT$ an ensemble set of 204 relaxed configurations were simulated and the respective values for criteria I,II were calculated (compare Table 3.2). In addition to the both criteria I, II the number of contacts of chromatin domains of two specific gene regions with the SC35 domains were also analyzed according to the experimental setup of Shopland et al. (2003).

The gene region 17q21 consists on ten chromatin domains and corresponds to a major part to a R-band and the 7p21 gene region consists on nine chromatin domains and the major part of this region is a G-band (compare Fig. 3.14).

The comparison of the calculated contact frequencies with the SCD model and the experimentally observed frequencies (compare Table 3.3) revealed a very good agreement for more than one contacts (between domains of the respective gene region and SC35 domains) at the same time for both gene regions. However there was no quantitative agreement for no or one contact, although qualitatively correctly reflected with the experiment (in the case of the gene region 17q21) was the behaviour that the frequency of no contacts is less than the number of at least one contact.

Table 3.2 Valuation of criteria I, II for the fitted optimal value $\varepsilon_{1,opt} = 1.90961^{+0.4248}_{-0.4616}kT$

Parameter	Experimentally observed (Shopland et al. 2003)	Calculated with the extended SCD model
G2RContacts	1/8 = 0.125	0.1190 ± 0.0290
contPerTer	3–4	3.4266 ± 0.2080

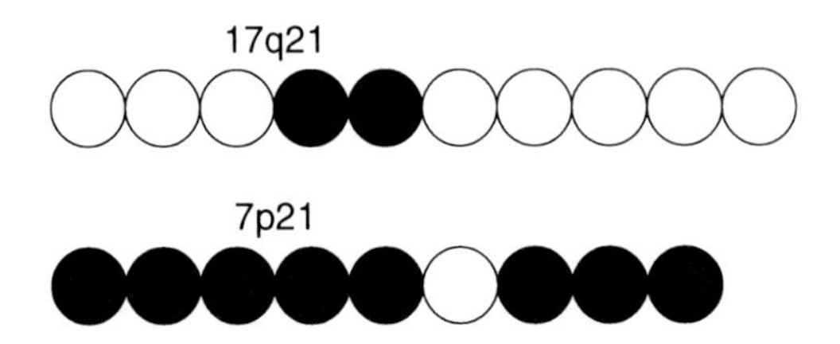

Fig. 3.14 Schematic drawing of the mapping of the gene regions 17q21 (*above*) and 7p21 (*down*) according to the SCD model. G-band domains are labelled in black and R-band domains in white

Table 3.3 Contact frequency of chromatin domains of the respective gene region with 0, 1, or >1 SC35 domains in 204 simulated cell nuclear configurations

		Percentage		
Gene region		0 contacts	1 contact	>1 contacts
7p21 (G)	Experimental	37 ± 4.79	52 ± 6	11 ± 2.24
	Calculated	56.86 ± 9.26	30.88 ± 6.05	12.25 ± 3.31
17q21 (R)	Experimental	1 ± 0.56	34 ± 4.29	63 ± 6.46
	Calculated	10.29 ± 2.97	22.06 ± 4.83	67.65 ± 10.49

On might argue that this out coming was a result of the rough estimation of the chromatin by spherical 1 Mbp chromatin domains (Table 3.3).

In another calculation (data are not shown here) possible rearrangements of R-bands to the CT surface and of G-band domains in the interior caused by the interactions with the SC35 domains were investigated. A cluster procedure identifying surface domains of a SCD-CT revealed that the presence of SC35 domains resulted in a different distribution of G-band and R-band domains at the surface. In contrast to a calculation where no SC35 domains were present, in the case with SC35 domains, more R-band domains are located at the surface and less were located in the interior. At the same time more G-band domains were located in the interior. This effect for the G-band domains was even stronger as for the R-band domains which might be the result of the attractive and repulsive interaction of R-band domains with SC35 domain while for G-band domains only repulsive interactions are present.

3.7 The Dynamics of Large Scale Nuclear Genome Structure in the Human Cell Nucleus

The nuclear architecture is not static but dynamic (Manders et al. 1999): Chromatin domains labeled directly with fluorochrome conjugated nucleotides revealed that these domains undergo constrained Brownian and occasionally also directed movements (Bornfleth et al. 1998; Bornfleth et al. 1999a,b; Schermelleh et al. 2001, Zink and Cremer 1998a; Zink et al. 1998b; Edelmann et al. 2001). Brownian motion of chromatin insensitive to metabolic inhibitors and constrained for each given segment to a subregion of the nucleus was also observed in Drosophila melanogaster nuclei (Marshall 1997). Therefore, the static biocomputing model approaches described above have to be extended to include also such movements.

The Monte Carlo method used above is based on stochastic methods and allows much faster calculations than the Brownian Dynamics method. With the Brownian Dynamics method, however, it is possible to predict the development in time. The Brownian Dynamics Simulations was based on the Langevin equation see (Mehring 1998).

$$m_i \frac{\partial^2 r_i}{\partial t^2} = -\sum_j \mu_{ij} \frac{\partial r_i}{\partial t} + F_i + \sum_j \alpha_{ij} f_j$$

F_i describes the interaction between the domains, r_i is the spatial coordinate, $\sum_j \propto_{ij} f_j$ is the term for the stochastical forces and the term for the friction is similar to $\sim\mu_{ij}$. This equation describes the movement of Brownian particles (the 1 Mbp domains) at constant temperature (Fig. 3.15).

Figure 3.16 presents sections through such nuclei simulated according to the Brownian motion 1 Mbp-SCD model. Figure 3.16c corresponds to a conventional epifluorescence image of the nucleus of a live lymphycyte cell with all CTs labelled, e.g., by labeling histones with a fluorescent protein (compare Gunkel et al. 2009).

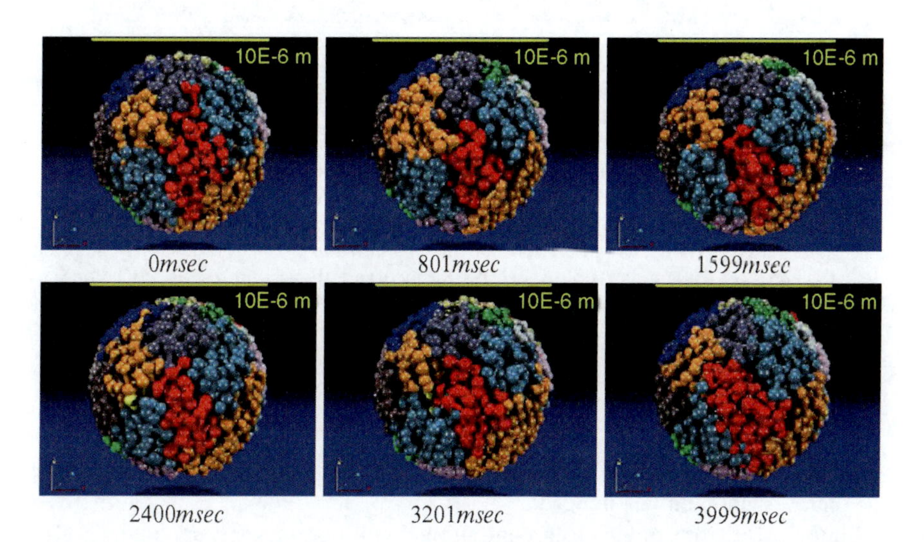

Fig. 3.15 Time series of the Brownian movement of chromosome territories of a simulated human lymphocyte nucleus according to the 1-Mbp SCD model. The starting point was a relaxed Monte Carlo simulation as described above. The temperature assumed was 310 K (37°C) and a viscosity of the medium of $\eta = 0.69 \times 10^{-3}$ kg/ms (corresponding to water); the total Brownian 1-Mbp-SCD model simulation time was 4 s; the individual timestep was 300 ns. The figure shows the configurations obtained for t = 0, 0.8, 1.6, 2.4, 3.2, and 4 s. The 1-Mbp domains of homologous CTs were labelled in the same color (From Kepper 2005)

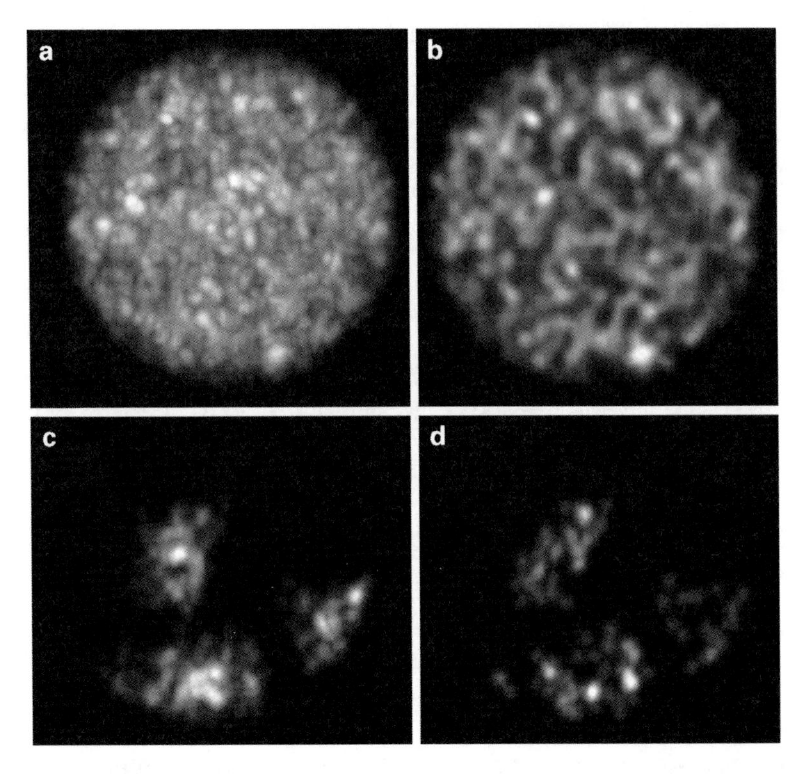

Fig. 3.16 Convolution of a Brownian dynamics simulation of an human nucleus with an experimental confocal point spread function (Bornfleth et al. 1998) to simulate the effects of 'blurring' due to the limited optical resolution (~250 nm laterally, 600 nm in the dirction of the optical axis) of a conventional confocal laser scanning fluorescence microscope (CLSM). Overall simulation time = 2 s; integration time step = 300 ns; imaging time step = 3 ms; temperature = 310 K; viscosity of the medium $\eta = 0.69 \times 10^{-3}$ kg/ms. (**a**) All CTs are shown *with* z-axis projection (i.e. a projection of all z-planes is visible);diameter of the nucleus: 10 μm; the same scale is used in (**b–d**); (**b**) all CTs are shown *without* z-axis projection (i.e. only one plane with a thickness of ~300 nm is visible); (**c**) CTs 18 and 19) are shown with z-axis projection (i.e. a projection of all planes is visible); (**d**) Centromere regions (simulated assuming a number of 1-Mbp domains in the region obtained from the centromere with z-axis projection (i.e. a projection of all planes is visible)

Figure 3.16b corresponds to a conventional CLSM image of such a nucleus where one optical section only is imaged.

Figure 3.16c corresponds to a conventional epifluorescence image where CTs #18 and 19 only have been visualized, and Fig. 3.16d corresponds to a conventional CLSM image where the centromeres only have been visualized. While such images are obtained readily using FISH of fixed cells, parts of individual CTs in vivo have been imaged using replication labelling techniques. The sites of centromeres have been labelled in vivo using Kinetochore specific proteins. Attempts to apply centromeric DNA-DNA hybridization schemes in vivo have also been described.

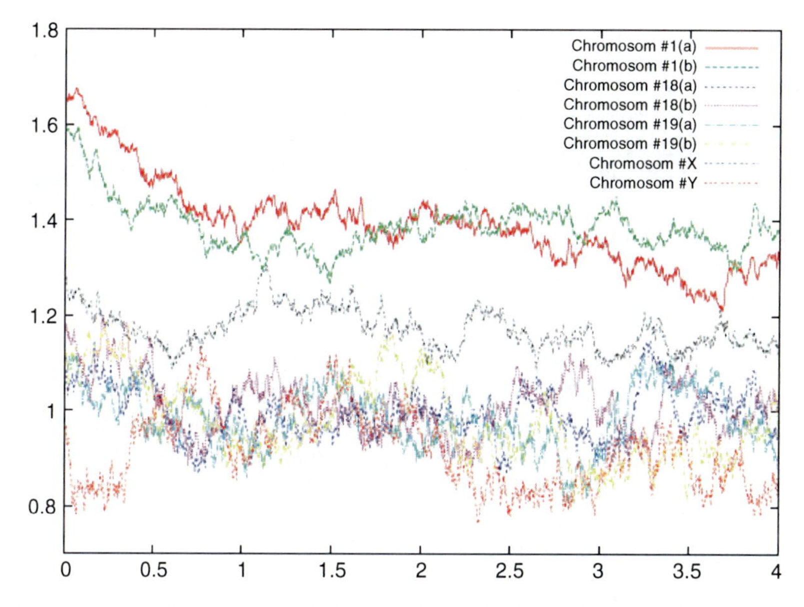

Fig. 3.17 *Ordinate*: Gyration radii of selected CTs simulated according to the Brownian motion 1-Mbp SCD model (for details see Fig. 3.13). *Abscissa*: Simulation time (s) from starting configuration (From Kepper 2005)

Figure 3.17 shows the gyration radii of selected CTs. The gyration radius is a measure for the spatial extension of a polymer chain, like the chain of 1-Mbp domains. It can be defined as

$$r_G^2 = \frac{1}{N}\sum_{i=1}^{N}\left\langle (r_i - r_C)^2 \right\rangle \qquad r_C = \frac{1}{N}\sum_{i=1}^{N} r_i$$

where r_i is the spatial coordinate (vector) of monomer (domain) I, and r_c is the coordinate (vector) of the geometrical barycentre. For example, a homogeneous sphere of radius r_k has a gyration radius of

$$r_G^2 = \frac{3}{5} r_K$$

For CTs with a large DNA content (such as the two homologous # a, 1b of human CT #1), the gyration radii are becoming smaller within the simulation time (total 4s). For the smaller CTs (#18,19, x,Y) the gyration radii appear to vary around a constant mean value.

Table 3.4 shows a result of such Brownian Dynamics simulations, starting with a relaxed Monte Carlo model as described in Section 3.2.

With the hardware used, the program takes about 1 month to finish 1 s of simulation. Faster network connections shorten the simulation times. In simulated nuclei each domain is clearly identified. For each time step the position of each domain is known. So it is possible to compute the distance between a domain in the telomere

Table 3.4 Simulated distances between telomere and centromere domains. A nucleus of a human cell simulated with a total time of 200 ms with Brownian Dynamics simulations; the integration time step was set to 50 ns, every 500 μs the positions of all domains were saved; the simulated temperature was 310 K, the viscosity $\eta = 0.69 \cdot 10^{-3}$ kg m·s (Chromosome sizes were taken from Morten 1991). Telomere and Centromere domains were identified according to the centromeric index

Chromosome	1	18	19	Y
First homologue	2.5 ± 0.2 μm	1.6 ± 0.5 μm	1.2 ± 0.3 μm	2.3 ± 0.2 μm
Second homologue	1.7 ± 0.2 μm	1.8 ± 0.2 μm	1.8 ± 0.4 μm	
Size in Mbp	263	98	85	59

and in the centromere region. To show the differences between the time steps the mean value and the standard deviation are calculated. In Table 3.4, the sizes for the chromosomes 1, 18, 19 and Y are given. The size of the chromosome does not correlate with the distance between the domain in the telomere and the centromere region. Centromere regions are not in the "middle" of chromosomes, there are two different arms with different lengths. The ratio differs with the chromosomes and is specific for the chromosome type.

3.8 Towards Quantitative Analysis of Nuclear Genome Nanostructure I: Computer Models

It is obvious that in real nuclei the 1 Mbp-domains of the SCD model cannot be spheres. This assumption simply means that the very complex 'nanostructure' of the chromatin inside the individual domains was not regarded. This deliberate oversimplification may be justified as long as the large scale nuclear genome architecture is considered, such as the general arrangement of chromosome territories in the nucleus, or even the mean distances between telomeres and centromere domains. However, spatial constraints of functional nuclear genome architecture, such as replication, transcription and repair, are intimately connected with the nanostructure, i.e. with dimensions considerably smaller than the 250 nm radius of a 1-Mbp-domain, any attempt to quantitative modelling such nanostructures has to be based on ideas on the nanoscale architecture of chromatin domains. In this respect, chromatin loops of different sizes have become a common textbook scheme to explain how chromatin packaging is achieved from the DNA level to the level of entire metaphase chromosomes. Such qualitative models need to be translated into quantitative, experimentally testable predictions; only in this way, the development of a consistent biophysical understanding of functional nuclear nanoarchitecture will become possible.

So far, several computer models have been proposed to numerically predict chromatin domain nanostructure up to the structure. For this, backfolding of chromatin fibres at some level is indispensable: In models that dismiss such backfolding, chromatin fibres expand throughout much of the nuclear space resulting in a non-territorial/non-domain interphase chromosome organization. In the random-walk/giant-loop (RW/GL) model, chromatin loops with a size of several megabases are backfolded to an underlying structure, but otherwise each giant loop is folded randomly. Another model, the multi-loop subcompartment (MLS) model, assumes that ~1-Mb chromatin domains are built up like a rosette from a series of chromatin-loop domains with sizes of ~100 kb, again assuming a random organization for each loop (Fig. 3.18). These models are compatible with the assumption that specific

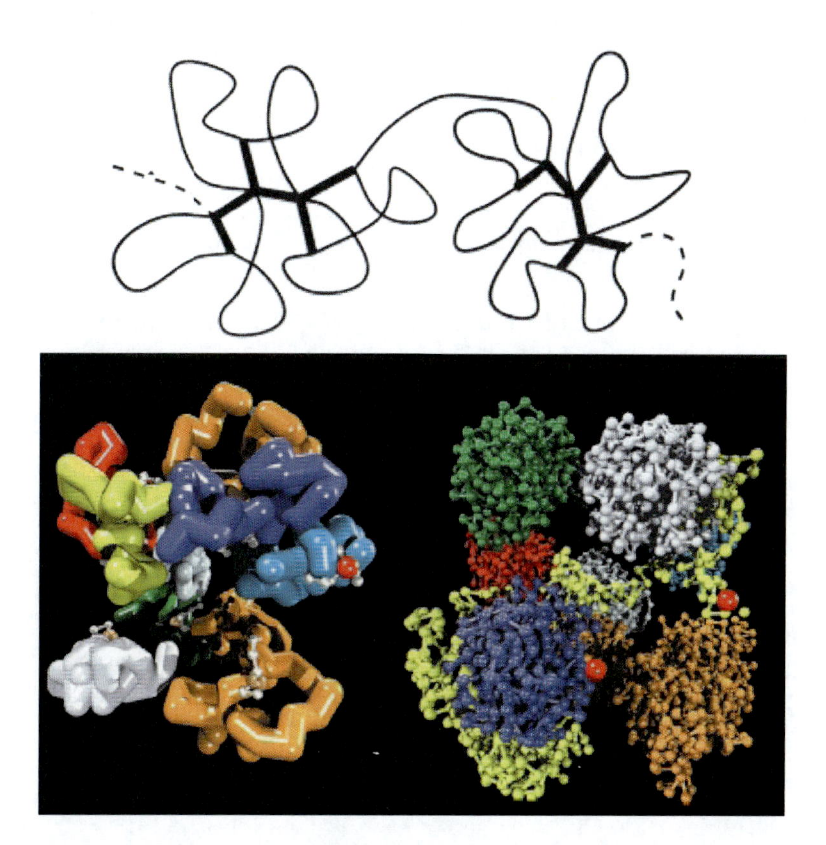

Fig. 3.18 The multiloop subcompartment model. (**a**) Two ~1-Mb chromatin domains or 'subcompartments' are shown linked by a chromatin fibre. Each ~1-Mb chromatin domain is built up as a rosette of looped ~100-kb chromatin fibres. At the centre, loops are held together by a magnified Loop base springsimulating the function of CT-Anchor proteins. (**b**, **c**) Two three-dimensional models of the internal ultrastructure of a ~1-Mb chromatin domain. (**b**) The nucleosome chain is compacted into a 30-nm chromatin fibre (visualized by cylinder segments) and folded into ten 100-kb-sized loop domains according to the multiloop subcompartment model. Occasionally, 30-nm fibres are interrupted by short regions of individual nucleosomes (*small white dots*). The arrow points to a red sphere, with a diameter of 30 nm, that represents a transcription factor complex. (**c**) Each of the ten 100-kb chromatin domains was modelled under the assumption of a restricted random walk (zig-zag) nucleosome chain. Each dot represents an individual nucleosome. Nine 100-kb chromatin domains are shown in a closed configuration and one in an open chromatin configuration with a relaxed chain structure that expands at the periphery of the 1-Mb domain. The open domain will have enhanced accessibility to partial transcription complexes preformed in the interchromatin compartment. By contrast, most of the chromatin in the nine closed domains remains inaccessible to larger factor complexes (From Cremer and Cremer 2001)

positions of genes inside or outside a chromatin-loop domain are not required for activation or silencing. However, they do make different predictions: first, about the extent of intermingling of different giant chromatin loops and chromatin-loop rosettes; and second, about the interphase distances between genes and other DNA segments that are located along a given chromosome.

Recently, single rosette-like parts of a domain were simulated and used for Brownian motion dynamics calculations to analyse the diffusion behaviour of small particles (corresponding to single proteins/protein complexes), and the accessibility of such particles in relation to the dynamic rosette structure (Odenheimer et al. 2009). Surprisingly, although the diffusion pattern of the diffusing particles revealed free diffusion, an area of about 6–12 kbp in the innermost part of these domains was revealed which is inaccessible even for small particles having the size of single proteins/protein complexes (Fig. 3.19). Three regions of different accessibility were obtained:

- An accessible region I (down to a distance of ~70 nm from the center of the rosette) where the density of a diffusing protein of ~50 kDa is reduced to ~50% of the freely accessible region (distance >180 nm)
- A region II difficult to access where the protein density drops to values around 10% of the freely accessible region

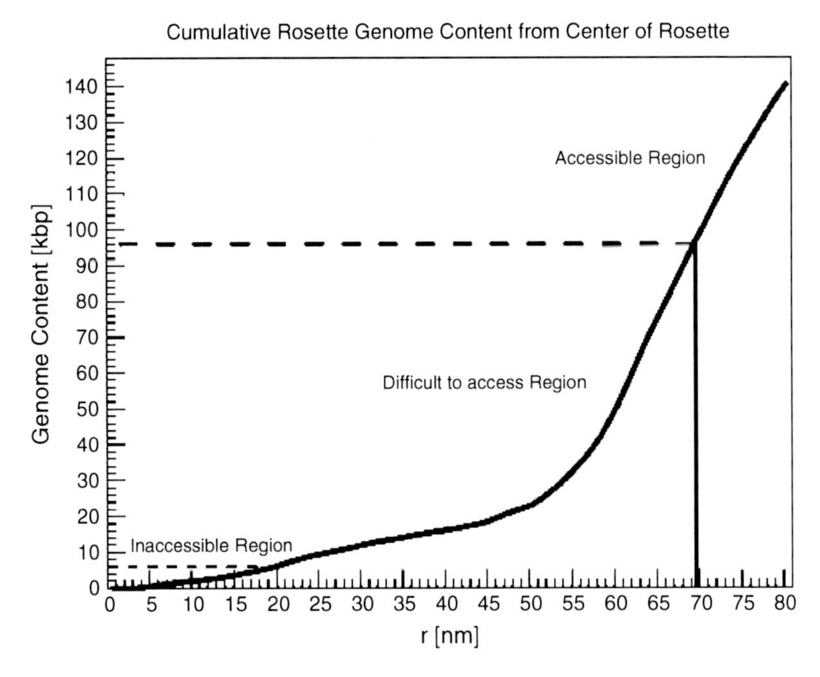

Fig. 3.19 Mbp-Nanostructure simulations are compatible with restricted accessibility of transcription factors. *Ordinate:* Genome content (kbp) with respect to the center of the Rosette nanostructure simulated (total length ~1 Mbp). *Abscissa*: Distance (nm) to the center of the 1-Mbp rosette (From Odenheimer et al. 2009)

– An in-accessible region III of about 3 kbp in length and a radius r_{Limit} = 20 nm where the relative protein density drops to very small densities (in the simulations to zero)

It is tempting to assume that a localisation of a promotor sequence in this area might silence the respective gene by the physical inaccessibility of this area for transcription factors: Thus, the RNA polymerase may not bind and the entire gene may not be transcribed. Thus, the silencing of genes in a chromatin domain could possibly be caused solely by physical inaccessibility. These predictions are well compatible with the findings of Verschure et al. (2003) based on conventional light microscopy observations.

3.9 Towards Quantitative Analysis of Nuclear Genome Nanostructure II: Perspectives of Superresolution Light Microscopy

As Fig. 3.20 indicates, in a conventional light microscopic observation volume corresponding to ~250 × 250 × 600 nm^3 , a region with a diameter of 2 × r_{Limit} = 2 × 20 nm = 40 nm was predicted to be inaccessible; this is only about 0.2% of the minimal conventional observation volume. Such a small region of inaccessibility is far below the detection limits of conventional light microscopy. Consequently, due to this lack of optical resolution the prediction of very small regions of inaccessibility in the range of few tens of nanometer is well compatible with the experimental results of tracking experiments of single streptavidin molecules in structurally and functionally distinct nuclear compartments. These conventional microscopy observations indicated that all nuclear subcompartments were easily and similarly accessible for such an average-sized protein, and even condensed heterochromatin neither excluded single molecules nor impeded their passage (Gruenwald et al. 2008). However, these protein molecules did not accumulate in heterochromatin, suggesting comparatively less free volume.

While the predictions of the nanostructure model presented are thus fully compatible with available conventional light microscopy data, a test of the predictions of this and other nanostructure models concerning small regions of inaccessibility requires microscopic techniques with an effective resolution in the range of few tens of nanometer. Until a few years ago, such a resolution was available only with electron microscopy (EM) approaches while for far field flight microscopy, such a resolution was regarded to be impossible to achieve.

Since the famous publication of Ernst Abbe (1873), this limit was regarded to be due to the wave nature of light and thus unsurmountable. Based on the diffraction theory of light, Ernst Abbe postulated for a specific color a “specific smallest distance which never can significantly surpass half a wavelength of blue light”, i.e. about 200 nm. A very similar conclusion was obtained in 1896 by Lord Rayleigh, “tracing the image representative of a mathematical point in the object, the point being regarded as self-luminous. The limit to definition depends upon the

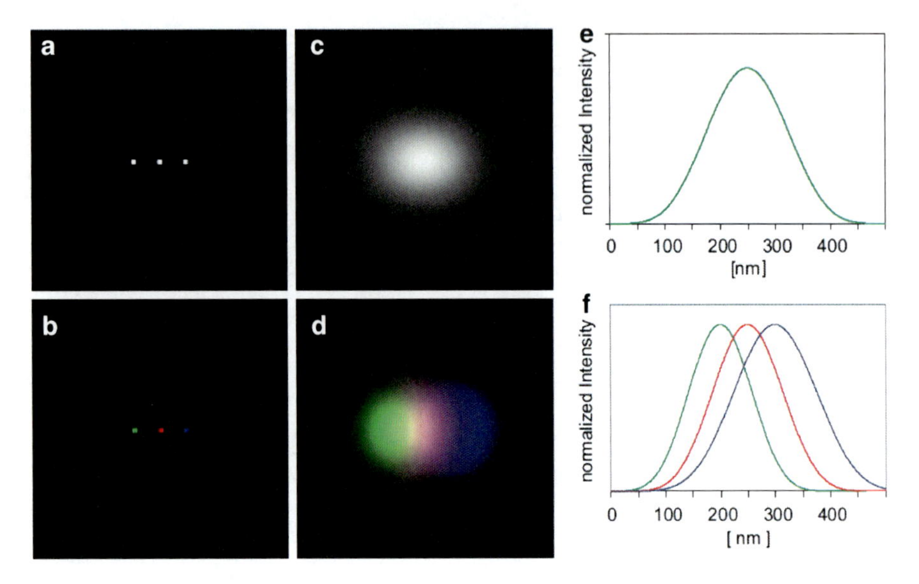

Fig. 3.20 Example for principle of spectral precision distance microscopy (SPDM). Three point-like objects are located within 50-nm distance of each other. The three point-like objects are labelled with the same spectral signature in (**a**) and with three different spectral signatures in (**b**). The computed system responses of the objects in (**a**) and (**b**) for a microscope with NA = 1.4/63 oil-immersion objective are shown in (**c**) and (**d**). Linescanes through the objects in (**c**) and (**d**) are shown in (**e**) and (**f**) respectively (From Cremer et al. 1999)

fact that owing to diffraction the image thrown even by a perfect lens ids not confined to a point, but distends itself over a patch or disk of light of finite diameter" Rayleigh (1896). From this, Rayleigh concluded "that the smallest resolvable distance ε is given by $\varepsilon = ½\ \lambda/\sin\alpha$, α being the wave-length in the medium where the object is situated, and α the divergence-angle of the extreme ray (the semi-angular aperture) in the same medium". Thus the same lightoptical (lateral) resolution limit of ca. 200 nm (for abbreviation often called 'Abbe-limit') is obtained in both theories.

Recently, however, a variety of laseroptical far field microscopy techniques based on fluorescence excitation has been developed to overcome the Abbe-limit of 200 nm at least in one direction, and to make possible Light Optical Analysis of BioStructures by Enhanced Resolution ("LOBSTER"). Some well known LOBSTER methods are confocal 4Pi-Laser Scanning Microscopy (Cremer and Cremer 1978; Hell and Wichmann 1994; Hell 2003; Egner et al. 2002, Bewersdorf et al. 2006, Baddeley et al. 2006, SMI microscopy (Hildenbrand et al. 2005; Martin et al. 2004) structured/patterned illumination microscopy (Heintzmann and Cremer 1999), STED microscopy (Hell and Wichmann 1994; Hell 2007; Schmidt et al. 2008; Donnert et al. 2006) or localization microscopy approaches using far field fluorescence microscopy (Cremer et al. 1996; Patwardhan 1997; Bornfleth et al. 1998; van Oijen et al. 1999; Edelmann et al. 1999; Esa et al. 2000; Lacoste et al. 2000; Schmidt et al. 2008; Heilemann et al. 2002; Betzig et al. 2006; Hess et al. 2006; Egner et al. 2002; Reymann et al. 2008; Lemmer et al. 2008). Using these latter techniques, an effective optical resolution in the 10–20 nm regime has been obtained. For the first time, this allowed "nanoimaging" of biostructures at macromolecular resolution using fluorescence excitation by visible light. While STED microscopy is a focused beam method allowing to rapidly image small regions of interest (few μm extension), the complementary "spectrally assigned localization microscopy" (SALM) techniques are preferentially used in non-focusing setups, thus allowing to rapidly image large regions of interest (50–100 μm extension).

The basis of SALM as a far field fluorescence microscopy "nanoimaging" approach using biocompatible temperature conditions is the independent localization of 'point like' objects excited to fluorescence emission by a focused laser beam or by non-focused illumination by 'optical isolation', i.e. the localization is assigned by appropriate spectral features. This approach, called by us 'spectral precision microscopy' (SPM) or spectral precision distance/position determination microscopy (SPDM) was conceived and realized in proof-of-principle experiments already in the 1990s (Cremer et al. 1996; Bornfleth et al. 1998; Cremer et al. 1999; Esa et al. 2000, 2001; Edelmann et al. 2000).

The principle of SPDM/SALM is shown schematically in Fig. 3.20. Let us assume three closely neighboring targets in a cell to be studied with distances much smaller than 1 FWHM, where FWHM represents the Full-Width-at-Half-Maximum of the effective Point Spread Function (PSF) of the optical system used. The 'point-like' (diameter much smaller than 1 FWHM) targets t_1, t_2 and t_3 (e.g., three molecules) are assumed to have been labeled with three different fluorescent spectral signatures specs1, specs2 and specs3. For example, t_1 is labeled with specs1, t_2 with specs2 and t_3 with specs3. The registration of the images (using focused or non-focused optical devices with a given FWHM) is performed in a spectrally discriminated way so that in a first image stack IM1, a specs1 intensity value I1 is assigned to each voxel v_k (k = 1,2,3,...) of the object space; in a second image stack IM2, a specs2 intensity value I2 is assigned to each voxel v_k of the object space; and in a third 3-D image stack IM3, a specs3 intensity value I3 is assigned to each voxel v_k of the object space. The positions of the objects/molecules obtained are assigned to a position map. Early 'proof-of-principle' experiments using confocal laser scanning fluorescence microscopy (CLSM) at room temperature to determine the positions (xyz) and mutual Euclidean distances of small sites on the same DNA molecule labelled with three different spectral signatures yielded a lateral effective resolution of about 30 nm (ca. 1/16th of the wavelength used) and ca. 50 nm axial effective resolution (Esa et al. 2000): Sites with a 3D distance of only 50 nm were still discriminated; their (xyz) positions were determined independently from each other with an error of few tens of nanometer, including the correction for optical aberrations. It was noted "that the SPM (Spectral Precision Microscopy = SPDM) strategy can be applied to more than two or three closely neighbored targets, if the neighboring targets t_1, t_2, t_3, ..., t_n have sufficiently different spectral signatures specs1, specs2,...specsn.... Three or more spectral signatures allow true structural conclusions.... Furthermore, it is clear that essentially the same SPM strategy can be applied also in all cases where the distance between targets of the *same* spectral signature is larger than FWHM. Computer simulations performed by (Bornfleth et al. 1998) indicated that a distance of about 1.5 FWHM is sufficient." (Cremer et al. 1999). Compared with related concepts (Betzig et al. 2006, van Oijen et al. 1999), in the SPDM approach the focus of application was to (a) to perform superresolution analysis by the direct evaluation of the positions of the spectrally separated objects; (b) to adapt the method to far field fluorescence microscopy at temperatures in the 300 K range ("room temperature"); (c) to specify spectral signatures to include all kinds of fluorescent emission parameters suitable, from absorption/emission spectra to fluorescence life times (Heilemann et al. 2002) to any other method allowing 'optical isolation' (Cremer et al. 2001 US Patent 7,298,461 B2) . Optical isolation means that at a given time and registration mode, the distance between two objects/molecules is larger than the FWHM. Early concepts included even the use of photoswitching to molecules from a state A to a state B in an observation volume in a reversible or irreversible way to improve the gain of nanostructural information (Cremer and Cremer 1972 DE).

In the last few years, approaches based on principles of SPDM and related methods ('spectrally assigned localization microscopy', SALM) have been considerably improved by several groups, especially by using photoswitchable fluorochromes. By imaging fluorescent bursts of single molecules after light activation, the position of the molecules was determined with a precision significantly better than the FWHM. These microscopic techniques were termed photoactivated localization microscopy (PALM) (Betzig et al. 2006), fluorescence photoactivation localization microscopy (FPALM) (Hess et al. 2006), stochastic optical reconstruction microscopy (STORM) (Rust et al. 2006) or PALM with independently running acquisition (PALMIRA) (Geisler et al. 2007, Andresen et al. 2008). In all these approaches, special fluorochromes were used which can be photoswitched between a 'dark' state A and a 'bright' state B (Hell Nature Biotech 2003).

Recently, we applied SPDM in the nanometer resolution scale to conventional fluorochromes that show "reversible photobleaching" or "blinking". "Reversible photobleaching" has been shown to be a general behavior in several fluorescent proteins, e.g., CFP, Citrine or eYFP (15–18). In such cases, the fluorescence emission of certain types can be described by assuming three different states of the molecule: A fluorescent state M_{fl}; a reversibly bleached state M_{rbl}; and an irreversibly bleached stat M_{ibl}. In the M_{rbl} state, the molecule can be excited to a large number of times to the M_{fl} state (fluorescence burst) until it passes into the irreversibly bleached state (i.e. a dark state long compared with the time required for a burst of fluorescence photons); and thus its position be determined. With velocity constants k_1, k_2, k_3 one can assume the transition scheme

$$M_{rbl} \underset{k_1}{\overset{k_2}{\rightleftarrows}} M_{fl} \xrightarrow{k_3} M_{ibl}$$

We recently showed that reversible photobleaching can be used for superresolution imaging ($SPDM_{Phymod}$) of cellular nanostructures labeled with conventional fluorochromes such as Alexa 488 (Reymann et al. 2008; Baddeley et al. 2009) or the Green Fluorescent Protein variant YFP (Lemmer et al. 2008). This novel extensions of the SPDM approach are based on the possibility to produce the optical isolation required by allowing only one molecule in a given observation volume to be in the M_{rbl}–M_{fl} cycle. In this case, each stochastic time onset ($t_1,t_2,t_3,\ldots t_k$) of a fluorescence burst is regarded to represent a specific PHYMOD spectral signature specs1,specs2,specs3,...specsk. Such spectral signatures characterized by the stochastic time onset of a emission of a fluorescent burst may be obtained by appropriate physical conditions, such as an exciting illumination intensity in the 10 kW/cm^2–1 MW/cm^2 range. We call fluorochromes with these characteristics "PHYMOD" (PHYsically MODifiable) fluorochromes. The spectral signatures realized in this way might be called "Phymod"specs. SPDM imaging appreoaches based on such "PHYMOD" Fluorochromes might be specified as $SPDM_{Phymod}$.

Figures 3.21, 3.22 and 3.23 show some recent examples how to use this superresolution technique for the lightoptical nanoscopy of nuclear genome nanoarchitecture (Fig. 3.24).

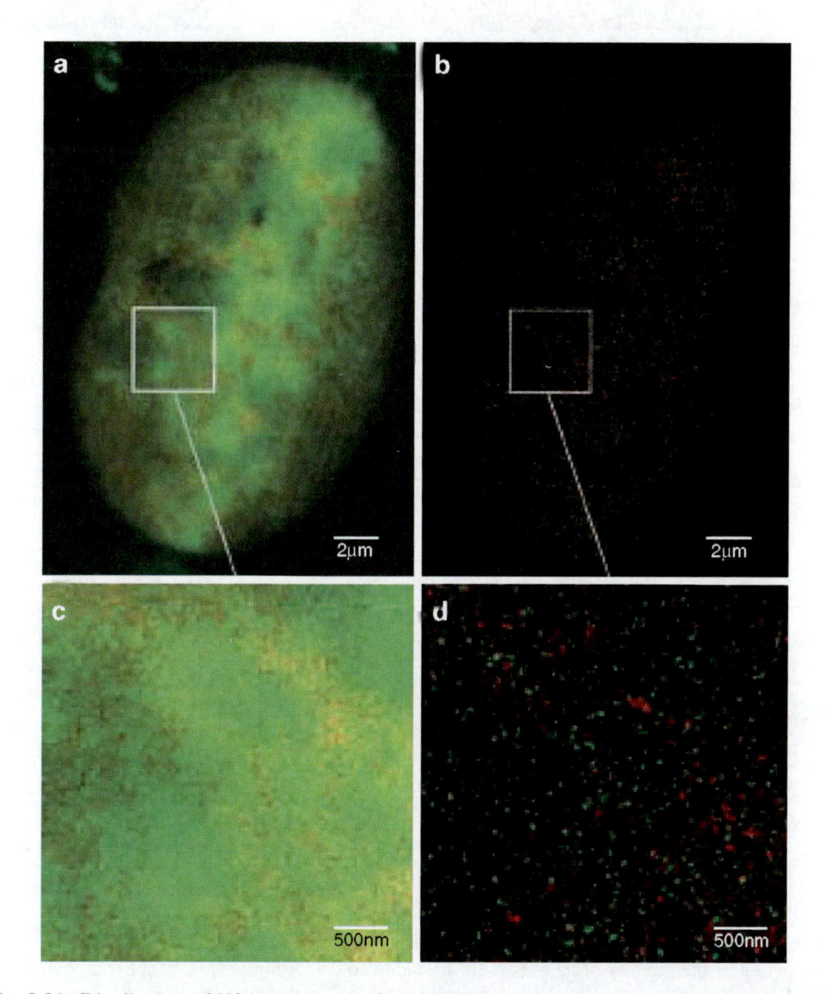

Fig. 3.21 Distribution of H2A proteins (*red*) and Snf2H proteins (*green*) within a human U2OS nucleus. (**a**) Conventional epifluorescence image. (**b**) 2CLM image, each fluorochrome-position is blurred with a Gaussian representing the individual localization accuracy. Panel (**c**) and (**d**) display enlarged regions. Scale bars are 2 μm in A and B and 500 nm in (**c**) and (**d**) (From Gunkel et al. 2009)

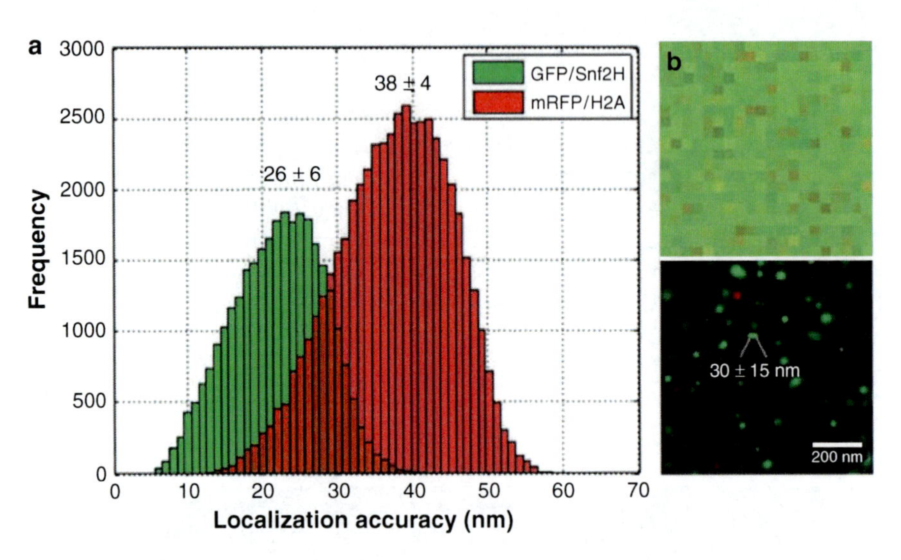

Fig. 3.22 Localization accuracy. (**a**) Histogram of localization accuracies of GFP and mRFP1 determined by the fitting algorithm for one cell (compare Fig. 3.20). (**b**) Comparison of conventional epifluorescence and 2CLM image. In the latter one, two molecule positions with a distance of about 30 nm can be resolved (both recorded within 7 s, corresponding to an estimated drift of about 1 nm). The pixel size in the epifluorescence image was 65 nm (From Gunkel et al. 2009)

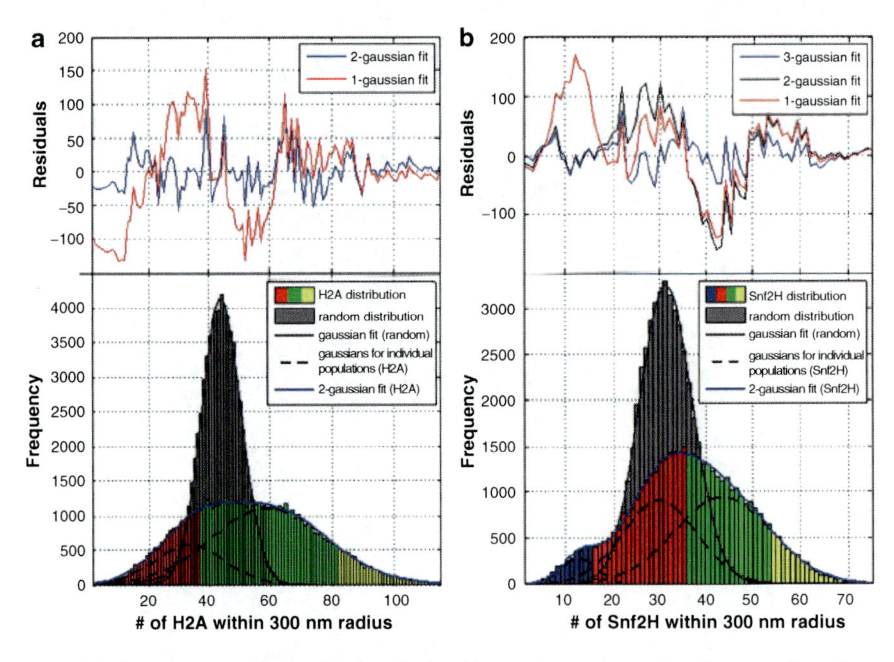

Fig. 3.23 Quantitative analysis of molecule densities on the nanoscale in a human cell nucleus. A histogram of the number of neighbors (of the same color) within a circle of 300 nm radius was plotted for H2A (**a**) and Snf2H (**b**) for the cell below. As a reference, the same histograms were plotted for random distributions with equal particle densities. Experimentally observed histograms were broader than the histograms for random distributions and could only be fitted considering different particle populations (From Gunkel et al. 2009)

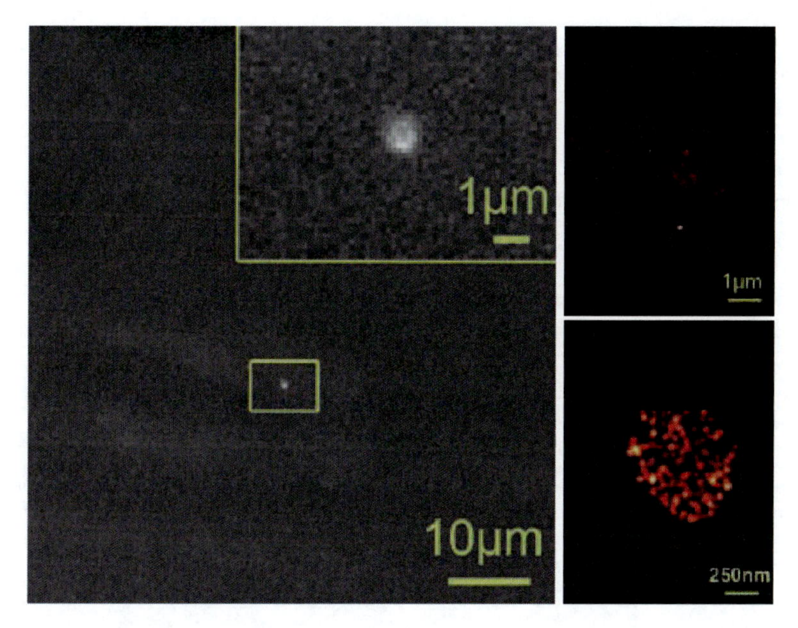

Fig. 3.24 Alexa 568 FISH labelled chromatin region in VH7 fibroblast cells. On the left the conventional wide-field fluorescence image and an enlarged insert of the heterochromatic target region on chromosome Y are shown (*upper middle*). On the upper right the corresponding SPDM image is displayed (From Lemmer et al. 2009)

3.9.1 Perspectives

Since its discovery in the 1840s of the nineteenth century, the biology of the nucleus has experienced an ever growing importance. With the advent of molecular cell biology, the nucleus as the site of the overwhelming amount of the cellular DNA has been studied very extensively. Presenly, the DNA sequence of the nuclei studied als well as the types of other molecules they contain is widely known. As a consequence, the focus is shifting to the influence of nuclear architecture on gene function. For example, for a long time most of the DNA in a human cell nucleus has been regarded to be just "junk" , because it did not contain RNA coding sequences. This concept is now changing: A large part of the non-coding DNA of the genome might have structural functions in the regulation of gene expression. To what extent this is the case is not known. However, more and more experimental evidence is accumulating for a major role of nuclear genome structure. In the present contribution, a number of examples has been given for the quantitative lightoptical analysis of nuclear architecture, from the 'macroscale' (chromosome territories and their positioning, down to 1 Mbp domains, several µm to ~0.2 µm) to first approaches to the "nanoscale" providing macromolecular lightoptical resolution (200–10 nm) of the positioning of small DNA repeats, individual histone molecules, and single chromatin remodelling factors. In addition, it was shown how structural information can be integrated into quantitative models of nuclear genome organisation and used to predict medically important consequences, such as the formation of cancer correlated chromosome translocations induced by ionizing radiation. It is anticipated that the new possibilties of lightoptical superresolution microscopy ('nanoscopy') and their correlation with ultrastructural methods, e.g., transmission electron microscopy (EM), Scanning EM (SEM), cryo-electron tomography, focused ion beam SEM, or X-ray microscopy will allow to gain nano-structural information until recently thought to be beyond the possibilities of experimental research. The integration of this knowledge into numerical models of nuclear structure down to the level of the nucleosomes will eventually make possible to develop a "virtual cell nucleus". This is expected to contribute substantially to a deeper insight into the basic understanding of life but also allow predictions valuable for cellular reprogramming and the effect of specific therapeutic agents.

Acknowledgements This work was supported by grants from the Deutsche Forschungsgemeinschaft and the European Union. For discussions and other help we especially thank Manuel Gunkel, Rainer Kaufmann and Thomas Cremer.

References

Abbe E (1873) Beitraege zur Theorie des Mikroskops und der mikroskopischen Wahrnehmung. Arch Mikrosk Anat 9:411–468

Albiez H, Cremer M, Tiberi C, Vecchio L, Schermelleh L, Dittrich S, Kuepper K, Joffe B, Thormeyer T, von Hase J, Yang S, Rohr K, Leonhardt H, Solovei I, Cremer C, Fakan S, Cremer T (2006) Chromatin domains and the interchromatin compartment form structurally defined and functionally interacting nuclear networks. Chromosome Res 14:707–733

Andresen M, Stiel AC, Folling J, Wenzel D, Schonle A, Egner A et al (2008) Photoswitchable fluorescent proteins enable monochromatic multilabel imaging and dual color fluorescence nanoscopy. Nat Biotechnol 26:1035–1040

Arsuaga J, Greulich-Bode KM, Vazquez M, Bruckner M, Hahnfeldt P, Brenner DJ, Sachs R, Hlatky L (2004) Chromosome spatial clustering inferred from radiogenic aberrations. Int J Radiat Biol 80:507–515

Baddeley D, Carl C, Cremer C (2006) 4Pi microscopy deconvolution with a variable point-spread function. Appl Opt 45:7056–7064

Baddeley D, Jayasinghe I, Cremer C, Cannell M, Soeller C (2009) Light-induced dark states of organic fluochromes enable 30 nm resolution imaging in standard media. Biophys J. doi:10.1016/j.bpj.2008.11.002

Ballarini F, Biaggi M, Ottolenghi A (2002) Nuclear architecture and radiation induced chromosome aberrations: models and simulations. Radiat Prot Dosimetry 99:175–182

Betzig E, Patterson GH, Sougrat R, Lindwasser OW, Olenych S, Bonifacino JS, Davidson MW, Lippincott-Schwartz J, Hess HF (2006) Imaging intracellular fluorescent proteins at nanometer resolution. Science 313:1642–1645

Bewersdorf J, Bennett BT, Knight KL (2006) H2AX chromatin structures and their response to DNA damage revealed by 4Pi microscopy. Proc Natl Acad Sci U S A 103:18137–18142

Bolzer A, Kreth G, Solovei I, Koehler D, Saracoglu K, Fauth Ch, Müller S, Eils R, Cremer C, Speicher M, Cremer T (2005) Three-dimensional maps of all chromosomes in human male fibroblast nuclei and prometaphase rosettes. PLOS Biol 3:e157

Bornfleth H, Sätzler K, Eils R, Cremer C (1998) High–precision distance measurements and volume–conserving segmenta tion of objects near and below the resolution limit in three–dimensional confoca l fluorescence microscopy. J Microsc 189:118–136

Bornfleth H, Edelmann P, Zink D, Cremer C (1999a) Handbook of computer vision and applications, vol III, Chapter Three – dimensional analysis of genome topology. Academic Press, San Diego/New York, pp 859–878

Bornfleth H, Edelmann P, Zink D, Cremer T, Cremer C (1999b) Quantitative motion analysis of sub–chromosomal foci in living cells using four–dimensional microscopy. Biophys J 75(5):2871–2886

Boyle S, Gilchrist S, Bridger JM, Mahy NL, Ellis JA, Bickmore WA (2001) The spatial organization of human chromosomes within the nuclei of normal and emerin-mutant cells. Hum Mol Genet 10:211–219

Branco MR, Pombo A (2007) Chromosome organization: new facts, new models. Trends Cell Biol

Bridger JML (1999) Analysis of mammalian interphase chromosome by FISH and immunofluorescence. In: Bickmore WA (ed) Chromsome structural ananlysis. Oxford University Press, Oxford, pp 103–123

Chevret E, Volpi EV, Sheer D (2000) Mini review: form and function in the human interphase chromosome. Cytogenetic Cell Genet 90(1–2):13–21

Cornforth MN, Greulich-Bode KM, Loucas BD, Arsuaga J, Vazquez M, Sachs RK, Bruckner M, Molls M, Hahnfeldt P, Hlatky L, Brenner DJ (2002) Chromosomes are predominantly located randomly with respect to each other in interphase human cells. J Cell Biol 159:237–244

Cremer C, Cremer T (1972) Verfahren zur Darstellung bzw. Modifikation von Objekt-Details, deren Abmessungen außerhalb der sichtbaren Wellenlänge liegen (Procedure to image or modify object details with dimensions below the wavelength of visible light). Ger Pat Appl P 21 16 521.9 (filed 5 April 1971)

Cremer C, Cremer T (1978) Considerations on a Laser-Scanning-Microscope with high resolution and depth of field. Microsc Acta 81:31–44

Cremer T (1985) Von der Zellenlehre zur Chromosomentheorie, Naturwissenschaftliche Erkenntnis und Theorienwechsel in der fr¨uhen Zell– und Vererbungsforschung. Springer, Heidelberg

Cremer T, Kurz A, Zirbel R, Dietzel S, Rinke B, Schröck E, Speicher MR, Mathieu U, Jauch A, Emmerich P, Scherthan H, Ried T, Cremer C, Lichter P (1993) Role of chromosome territories in the functional compartmentalization of the cell nucleus. Cold Spring Harb Symp Quant Biol 85:777–792

Cremer T, Lichter P, Borden J, Ward DC, Manuelidis L (1988) Detection of chromosome aberrations in metaphase and interphase tumor cells by in situ hybridization using chromosome–specific library probes. Hum Genet 80:235–246

Cremer T, Kurz A, Zirbel R, Dietzel S, Rinke B, Schröck E, et al. (1993) Role of chromosome territories in the functional compartmentalization of the cell nucleas. Cold Spring Harb Symp Quant Biol. 1993;85:777–792

Cremer C, Muenkel C, Granzow M, Jauch A, Dietzel S, Eils R, Guan XY, Meltzer PS, Trent JM, Langowski J, Cremer T (1996) Nuclear architecture and the induction of chromosomal aberrations. Mutat Res 366:97–116

Cremer C, Edelmann P, Bornfleth H, Kreth G, Muench H, Luz H, Hausmann M (1999) Principles of Spectral Precision Distance confocal microscopy for the analysis of molecular nuclear structure. In: Jähne B, Haußecker H, Geißler P (eds) Handbook of computer vision and applications, vol 3. Academic Press, San Diego/New York, pp 839–857

Cremer T, Kreth G, Koester H, Fink RH, Heintzmann R, Cremer M, Solovei I, Zink D, Cremer C (2000) Chromosome territories, interchromatin domain compartment, and nuclear matrix: an integrated view of the functional nuclear architecture. Crit R Eukaryot Gene Express 12:179–212

Cremer M, Hase JV, Volm T, Brero A, Kreth G, Walter J, Fischer C, Solovei I, Cremer C, Cremer T (2001) Non-random radial higher-order chromatin arrangements in nuclei of diploid human cells. Chrom Res 9:541–567

Cremer T, Cremer C (2001) Chromosome territiories, nuclear architecture and gene regulation in mammalian cells. Nat Rev Genet 2(4):292–301

Cremer M, Kupper K, Wagler B, Wizelman L, von Hase J, Weiland Y, Kreja L, Diebold J, Speicher MR, Cremer T (2003) Inheritance of gene density-related higher order chromatin arrangements in normal and tumor cell nuclei. J Cell Biol 162:809–820

Cremer T, Cremer M, Dietzel S, Mueller S, Solovei I, Fakan S (2006) Chromosome territories – a functional nuclear landscape. Curr Opin Cell Biol 18:307–316

Cremer T, Cremer C (2006a) Rise, fall and resurrection of chromosome territories: a historical perspective. Part I. The rise of chromosome territories. Eur J Histochem 50:161–176

Cremer T, Cremer C (2006b) Rise, fall and resurrection of chromosome territories: a historical perspective. Part II. Fall and resurrection of chromosome territories during the 1950s to 1980s. Part III. Chromosome territories and the functional nuclear architecture: experiments and models from the 1990s to the present. Eur J Histochem 50:223–272

Croft JA, Bridger JM, Boyle S, Perry P, Teague P, Bickmore WA (1999) Differences in the localisation and morphology of chromosomes in the human nucleus. J Cell Biol 145:1119–1131

Dietzel S, Eils R, Saetzler K, Bornfleth H, Jauch A, Cremer C, Cremer T (1998) Evidence against a looped structure of the inactive human X–chromosome territory. Exp Cell Res 240:187–196

Dimitrova DS, Berezney R (2002) The spatio-temporal organization of DNA replication sites is identical in primary, immortalized and transformed mammalian cells. J Cell Sci 115:4037–4051

Donnert G, Keller J, Medda R, Andrei MA, Rizzoli S, Luhrmann R, Jahn R, Eggeling C, Hell SW (2006) Macromolecular–scale resolution in biological fluorescence microscopy. Proc Natl Acad Sci U S A 103:11440–11445

Deloukas P, Schuler GD, Gyapay G, Beasley EM, Soderlund C, Rodriguez-Tomé P, Hui L, Matise TC, McKusick KB, Beckmann JS, Bentolila S, Bihoreau M, Birren BB, Browne J, Butler A, Castle AB, Chiannilkulchai N, Clee C, Day PJ, Dehejia A, Dibling T, Drouot N, Duprat S, Fizames C, Fox S, Gelling S, Green L, Harrison P, Hocking R, Holloway E, Hunt S, Keil S, Lijnzaad P, Louis-Dit-Sully C, Ma J, Mendis A, Miller J, Morissette J, Muselet D, Nusbaum HC, Peck A, Rozen S, Simon D, Slonim DK, Staples R, Stein LD, Stewart EA, Suchard MA, Thangarajah T, Vega-Czarny N, Webber C, Wu X, Hudson J, Auffray C, Nomura N, Sikela JM, Polymeropoulos MH, James MR, Lander ES, Hudson TJ, Myers RM, Cox DR, Weissenbach J, Boguski MS, Bentley DR (1998) A physical map of 30, 000 human genes. Science 282:744–746

Dundr M, Misteli T (2001) Functional architecture in the cell nucleus. Biochem J 356:297–310

Edelmann P, Bornfleth H, Zink D, Cremer T, Cremer C (2001) Morphology and dynamics of chromosome territories in living cells. Biochimica et Biophysica Acta 1551:M29–M40

Edelmann P, Cremer C (2000) Improvement of confocal spectral precision distance microscopy (SPDM). In: Farkas DL, Leif RC (eds) Optical diagnostics of Living Cells III. Proceedings of SPIE, vol 3921, pp 313–320

Edelmann P, Esa A, Hausmann M, Cremer C (1999) Confocal laser-scanning fluorescence microscopy: in situ determination of the confocal point-spread function and the chromatic shifts in intact cell nuclei. Optik 110:194–198

Edwards AA (1997) The use of chromosomal aberrations in human lymphocytes for biological dosimetry. Radiat Res 148:539–544

Egner A, Jakobs S, Hell SW (2002) Fast 100-nm resolution three-dimensional microscope reveals structural plasticity of mitochondria in live yeast. Proc Natl Acad Sci U S A 99:3370–3375

Esa A, Edelmann P, Trakthenbrot L, Amariglio N, Rechavi G, Hausmann M, Cremer C (2000) 3D–spectral precision distance microscopy (SPDM) of chromatin nanostructures after triple–colour labeling: a study of the BCR region on chromosome 22 and the Philadelphia chromosome. J Microsc 199:96–105

Esa A, Coleman AE, Edelmann P, Silva S, Cremer C, Janz S (2001) Conformational differences in the 3D-nanostructure of the immunoglobulin heavy-chain locus, a hotspot of chromosomal translocations in B lymphocytes. Cancer Genet Cytogen 127:168–173

Ferreira J, Paolella G, Ramos C, Lamond AI (1997) Spatial organization of large-scale chromatin domains in the nucleus: a magnified view of single chromosme territories. J Cell Biol 139:1597–1610

Friedland W, Paretzke HG, Ballarini F, Ottolenghi A, Kreth G, Cremer C (2008) First steps towards systems radiation biology studies concerned with DNA and chromosome structure within living cells. Radiat Environ Biophys 47:49–61

Geisler P, Schoenle A, von Middendorff C, Boch H, Eggeling C, Egner A et al (2007) Resolution of l/10 in fluorescence microscopy using fast single molecule photoswitching. Appl Phys A 88:223–226

Gonzalez-Melendi P, Beven A, Boudonck K, Abranches R, Wells B, Dolan L, Shaw P (2000) The nucleus: a highly organized but dynamic structure. J Microsc 198:199–207

Gruenwald D, Martin RM, Buschmann V, Bazett-Jones DP, Leonhardt H, Kubitscheck U, Cardoso MC (2008) Probing intranuclear environments at the single-molecule level. Biophys J 94:2847–2858

Gunkel M, Erdel F, Rippe K, Lemmer P, Kaufmann K, Hoermann C, Amberger R, Cremer C (2009) Dual color localization microscopy of cellular nanostructures, Biotechnology J: S927–938

Heilemann M, Herten DP, Heintzmann R, Cremer C, Muller C, Tinnefeld P, Weston KD, Wolfrum J, Sauer M (2002) High–resolution colocalization of single dye molecules by fluorescence lifetime imaging microscopy. Anal Chem 74:3511–3517

Heintzmann R, Cremer C (1999) Lateral modulated excitation microscopy: improvement of resolution by using a diffraction grating. SPIE 3568:185–196

Hell SW (2003) Toward fluorescence nanoscopy. Nat Biotechnol 21:1347–1355

Hell SW (2007) Far-field optical nanoscopy. Science 316:1153–1158

Hell SW, Wichmann J (1994) Breaking the diffraction resolution limit by stimulated emission: stimulated-emission-depletion fluorescence microscopy. Opt Lett 19:780–782

Hess ST, Girirajan TP, Mason MD (2006) Ultra-high resolution imaging by fluorescence photoactivation localization microscopy. Biophys J 91:4258–4272

Hildenbrand G, Rapp A, Spoeri U, Wagner C, Cremer C, Hausmann M (2005) Nano-sizing of specific gene domains in intact human cell nuclei by spatially modulated illumination light microscopy. Biophys J 88:4312–4318

Holley WR, Mian IS, Park SJ, Rydberg B, Chatterjee A (2002) A model for interphase chromosomes and evaluation of radiation-induced aberrations. Radiat Res 158:568–580

International Human Genome Sequencing Consortium (2004) Finishing the euchromatic sequence of the human genome. Nature 431:931–945

Johnston PJ, Olive PL, Bryant PE (1997) Higher-order chromatin structure-dependent repair of DNA double-strand breaks: modelling the elution of DNA from nucleoids. Radiat Res 148:561–567

Kepper N (2005) Brownsche Dynamik Simulationen von ganzen Zellkernen zur Bestimmung der Bewegungsmuster von Chromosomen und Genregionen auf der Grundlage des 1-Mbp SCD Modells. Dissertation, University Heidelberg

Kepper N, Foethke D, Stehr R, Wedeman G, Rippe K (2008) Nucleosome geometry and internucleosomal interactions control the chromatin fiber conformation. Biophys J 95:3692–3705

Kozubek S, Lukasova E, Mareckova A, Skalnikova M, Kozubek M, Bartova E, Kroha V, Krahulcova E, Slotova J (1999) The topological organization of chromosomes 9 and 22 in cell nuclei has a determinative role in the induction of t(9, 22) translocations and in the pathogenesis of t(9, 22) leukemias. Chromosoma 108:426–435

Kreth G, Finsterle J, von Hase J, Cremer M, Cremer C (2004) Radial arrangement of chromosome territories in human cell nuclei: a computer model approach based on gene density indicates a probabilistic global positioning code. Biophys J 86:2803–2812

Kreth GJ, Finsterle CC (2004b) Virtual radiation biophysics: implications of nuclear structure. Cytogenet Genome Res 104:157–161

Kreth G, Pazhanisamy SK, Hausmann M, Cremer C (2007) Cell type specific quantitative predictions of radiation-induced chromosomal aberrations by a computer model approach. Radiat Res 167:515–525

Lacoste TD, Michalet X, Pinaud F, Chemla DS, Alivisatos AP, Weiss S (2000) Ultrahigh-resolution multicolor colocalization of single fluorescent probes. Proc Natl Acad Sci U S A 97:9461–9466

Lemmer P, Gunkel M, Baddeley D, Kaufmann R, Urich A, Weiland Y, Reymann J, Müller P, Hausmann M, Cremer C (2008) SPDM: light microscopy with single-molecule resolution at the nanoscale. Appl Phys B Lasers O. doi:10.1007/s00340-008-3152-x

Lemmer P, Gunkel M, Weiland Y, Mueller P, Baddeley D, Kaufmann R, Urich A, Eipel H, Amberger R, Hausmann M, Cremer C (2009) Using conventional fluorescent markers for far-field fluorescence localization nanoscopy allows resolution in the 10-nm range. J Microsc 235:S163–171

Lamond AI, Earnshaw WC (1998) Structure and function in the nucleus. Science 280:547–553

Leitch AR (Mar 2000) Higher levels of organization in interphase nucleus of cycling and differentiated cells. Microbiol Mol Bio Rev 64(1):138–152

Lloyd DC, Edwards AA, Prosser JS (1986) Chromosome aberrations induced in human lymphocytes by in vitro acute X and gamma radiation. Radiat. Prot. Dosim. 15, 83–88

Manders EM, Kimura H, Cook PR (1999) Direct imaging of DNA in living cells reveals the dynamics of chromosome formation. J Cell Biol 144:813–821

Marshall WF, Straight A, Marko JF, Swedlow J, Dernburg A, Belmont A, Murray AW, Agard DA, Sedat JW (1997) Interphase chromosomes undergo constrained diffusional motion in living cells. Curr Biol 7:930–939

Martin S, Failla AV, Spöri U, Cremer C, Pombo A (2004) Measuring the size of biological nanostructures with spatially modulated illumination microscopy. Mol Biol Cell 15:2449–2455

Mehring C (1998) Simulation of chromosomes. Master's thesis, Ruperto-Carola University of Heidelberg, Germany

Metropolis N, Rosenbluth AW, Rosenbluth MN, Teller AH, Teller E (1953) Equation of state calculations by fast computing machines. J Chem Phys 21:1087–1092

Morten NE (1991) Parameters of the human genome. Proc Natl Acad Sci U S A 88:7474–7476

Muenkel C, Langowski J (1998) Chromosome structure predicted by a polymer model. Phys Rev E 57:5888–5896

Nicodemi M, Prisco A (2009) Thermodynamic pathways to genome spatial organization in the cell nucleus. Biophys J 96:2168–2177

O'Brien TP, Bult CJ, Cremer C, Grunze M, Knowles BB, Langowski J, McNally J, Pederson T, Politz JC, Pombo A, Schmahl G, Spatz JP, van Driel R (2003) Genome function and nuclear architecture: from gene expression to nanoscience. Genome Res 13(6):1029–1041

Odenheimer J, Heermann D, Kreth G (2009) Brownian dynamics simulations reveal regulatory properties of higher-order chromatin structures. Euro Biophys J 38(6):749–756

Parada L, Misteli T (2002) Chromosome positioning in the interphase nucleus. Trends Cell Biol 12(9):425–432

Parada L, Sotiriou S, Misteli T (2004) Spatial genome organization. Exp Cell Res 296:64–70

Patwardhan A (1997) Subpixel position measurement using 1D, 2D, 3D centroid algorithms with emphasis on applications in confocal microscopy. J Microsc 186:246–257

Rippe K, Mazurkiewicz J, Kepper N (2008) Interactions of histones with DNA: nucleosome assembly, stability and dynamics. In: Dias RS, Lindman B (eds) DNA interactions with polymers and surfactants. Wiley, London, pp 135–172. ISBN 978-0-8138-0646-4

Reymann J, Baddeley D, Gunkel M, Lemmer P, Stadter W, Jegou T, Rippe K, Cremer C, Birk U (2008) High-precision structural analysis of subnuclear complexes in fixed and live cells via spatially modulated illumination (SMI) microscopy. Chromosome R 16:367–382

Roix JJ, McQueen PG, Munson PJ, Parada LA, Misteli T (2003) Spatial proximity of translocation-prone gene loci in human lymphomas. Nat Genet 34:287–291

Rust MJ, Bates M, Zhuang X (2006) Sub-diffraction-limit imaging by stochastic optical reconstruction microscopy (STORM). Nat Meth 3:793–795

Qumsiyeh MB (Jul 1999) Structure and function of the nucleus: anatomy and physiology of chromatin. Cell Mol Life Sci 55(8–9):1129–1140

Sachs RK, Chen AM, Brenner DJ (1997) Proximity effects in the production of chromosome aberrations by ionizing radiation. Int J Radiat Biol 71:1–19

Sachs RK, Levy D, Chen AM, Simpson PJ, Cornforth MN, Ingerman E, Hahnfeldt P, Hlatky LR (2000) Random breakage-and reunion chromosome aberration formation model: an interaction-distance implementation based on chromatin geometry. Int J Radiat Biol 76:1579–1588

Schermelleh L, Solovei I, Zink D, Cremer T (2001) Two-color fluorescence labelling of early and mid–to–late replicating chromatin in living cells. Chromosome R 9:77–80

Schmidt R, Wurm CA, Jakobs S, Engelhardt J, Egner A, Hell SW (2008) Spherical nanosized focal spot unravels the interior of cells. Nat Meth 5:539–544

Shopland L, Johnson C, Byron M, McNeil J, Lawrence J (2003) R-bands around SC-35 domains: evidence for local euchromatic neighbourhoods. J Cell Biol 162(6):981–990

Schroeck E, du Manoir S, Veldman T, Schoell B, Wienberg J, Ferguson-Smith MA, Ning Y, Ledbetter DH, Bar-Am I, Soenksen D, Garini Y, Ried T (1996) Multicolour spectral karyotyping of human chromosomes. Science 273:494–497

Solovei I, Cavallo A, Schermelleh L, Jaunin F, Scasselati C, Cmarko D, Cremer C, Fakan S, Cremer T (2002) Spatial preservation of nuclear chromatin architecture during three-dimensional fluorescence in situ hybridization (3D–FISH). Exp Cell Res 276:10–23

Speicher MR, Ballard SG, Ward DC (1996) Karyotyping human chromosomes by combinatorial multi-fluor FISH. Nat Genet 12:368–375

Sun HB, Shen J, Yokota H (2000) Size-dependent positioning of human chromosomes in interphase nuclei. Biophys J 79:184–190

Tanabe H, Müller St, Neusser M, von Hase J, Calcagno E, Cremer M, Solovei I, Cremer C, Cremer T (2002) Evolutionary conservation of chromosome territory arrangements in cell nuclei from higher primates. P Nat Acad Sci U S A 99:4424–4429

Tsukamoto T, Hashiguchi N, Janicki SM, Tumbar T, Belmont AS, Spector DL (2000) Visualization of gene activity in living cells. Nat Cell Biol 2(12):871–878

van Oijen AM, Koehler J, Schmidt J, Muller M, Brakenhoff GJ (1999) Far-field fluorescence microscopy beyond the diffraction limit. J Opt Soc Am A 16:909–991

Verschure PJ, van der Kraan I et al (2003) Condensed chromatin domains in the mammalian nucleus are accessible to large macromolecules. EMBO reports 4(9):861–866

Visser AE, Eils R, Jauch A, Little G, Bakker PJM, Cremer T, Aten JA (1998) Spatial distribution of early and late replicating chromatin in interphase chromosome territories. Exp Cell Res 243:398–407

Visser AE, Jaunin F, Fakan S, Aten JA (2000) High resolution analysis of interphase chromosome domains. J Cell Sci 113(Pt 14):2585–2593

Weierich C, Brero A, Stein S, Hase JV, Cremer C, Cremer T, Solovei I (2003) Three-dimensional arrangements of centromeres and telomeres in nuclei of human and murine lymphocytes. Chrom Res 11(5):485–502

Xing Y, Johnson CV, Moen PTJ, McNeil JA, Lawrence J (1995) Nonrandom gene organization: structural arrangements of specific pre-mRNA transcription and splicing with SC-35 domains. J Cell Biol 131:1635–1647

Zink D, Bornfleth H, Visser A, Cremer C, Cremer T (1999) Organization of early and late replicating DNA in human chromosome territories. Exp Cell Res 247:176–188

Zink D, Cremer T (1998) Cell nucleus: chromosome dynamics in nuclei of living cells. Curr Biol 8:R321–324

Zink D, Cremer T, Saffrich R, Fischer R, Trendelenburg MF, Ansorge W, Stelzer EH (1998) Structure and dynamics of human interphase chromosome territories in vivo. Hum Genet 102(2):241–251

Zirbel RM, Mathieu UR, Kurz A, Cremer T, Lichter P (1993) Evidence for a nuclear compartment of transcription and splicing located at chromosome domain boundaries. Chromosome Res 1(2):93–106

Zuckerkandl E (1997) Junk DNA and sectorial gene repression. Gene 205(1–2):323–343

Chapter 4
Statistical Shape Theory and Registration Methods for Analyzing the 3D Architecture of Chromatin in Interphase Cell Nuclei

Siwei Yang, Doris Illner, Kathrin Teller, Irina Solovei, Roel van Driel, Boris Joffe, Thomas Cremer, Roland Eils, and Karl Rohr

Abstract For studying the 3D structure formed by consecutive genomic regions of chromosomes we propose using statistical shape theory in conjunction with registration methods. In contrast to earlier work, where the 3D chromatin structure was analyzed indirectly, we here directly exploit the 3D locations of genomic regions to determine the large-scale structure of chromatin fiber. Our study is based on 3D microscopy images of the X-chromosome where four consecutive genomic regions (BACs) have been simultaneously labeled by multicolor FISH. To allow unique reconstruction of the 3D shape, image data with sets of four consecutive BACs have been acquired with an overlap of three BACs between the different sets. We have statistically analyzed the data and it turned out that for all datasets the 3D structure is non-random. In addition, we found that the shapes of active and inactive X-chromosomal genomic regions are statistically independent. Moreover, we reconstructed the 3D structure of chromatin in a small genomic region based on five BACs resulting from two overlapping four BACs. We also found that spatial normalization of cell nuclei using non-rigid image registration has a significant influence on the location of the genomic regions.

K. Rohr (✉), S. Yang, and R. Eils
Department of Bioinformatics and Functional Genomics, Biomedical Computer Vision Group (BMCV), University of Heidelberg, BioQuant Center, IPMB, and German Cancer Research Center (DKFZ), Im Neuenheimer Feld 267, 69120, Heidelberg, Germany
e-mail: K.Rohr@dkfz-heidelberg.de

D. Illner, K. Teller, I. Solovei, B. Joffe, and T. Cremer
Department of Biology II, Anthropology and Human Genetics, Ludwig Maximilians University Munich, Biozentrum, Germany

R. van Driel
Research group: Structural and Functional Organization of the Cell Nucleus, Swammerdam Institute for Life Sciences (SILS), University of Amsterdam, BioCentrum Amsterdam, Kruislaan 318, 1098 SM Amsterdam, The Netherlands

N.M. Adams and P.S. Freemont (eds.), *Advances in Nuclear Architecture*,
DOI 10.1007/978-90-481-9899-3_4,

4.1 Introduction

The 3D architecture of chromatin formed by different genomic regions is still poorly understood although this spatial information is essential for understanding cellular function. In particular, the positions of genes within the cell nucleus play an important role in transcriptional regulation (see Fraser and Bickmore 2007; Lanctôt et al. 2007; Sexton et al. 2007; Soutoglou and Misteli 2007 for reviews). For example, some genes are known to move from the nuclear envelope upon transcriptional activation and towards the envelope when silenced (e.g., Chuang et al. 2006; Chuang and Belmont 2007; Dundr et al. 2007; Kosak and Groudine 2004; Ragoczy et al. 2006).

In this work, we study the 3D architecture of interphase chromatin within a small region of the X-chromosome. The X-chromosome is one of the two sex-determining chromosomes in many animal species. Early in the embryonic development of females, one of the two X-chromosomes is inactivated in almost all somatic cells. X-inactivation ensures that the female, with two X-chromosomes, does not have twice as many X-chromosomal gene products as the male, which only has a single X-chromosome (Lyon 1961). The folding structure of X-chromatin is still poorly understood, despite the fact that the 3D organization of the chromatin plays an important role in the control of gene expression (Sproul et al. 2005). The aim of our work is threefold: First, we examine whether the large scale folding structure of the X-chromatin is non-random. Second, we study the relationship between the folding structures of active X-chromatin (Xa) and inactive X-chromatin (Xi). Third, we reconstruct and analyze the 3D structure formed by consecutive genomic regions.

In previous work, 3D chromatin structure was analyzed *indirectly*, for example, by measuring physical distances (e.g., Yokota et al. 1995; Ostashevsky 1998; Skalníková et al. 2000; Parreira et al. 1997) or angles (e.g., Zlatanova et al. 1998; Kozubek et al. 2002), which have been computed based on the positions of labelled genomic regions. Also, the centers of nuclei have been used to determine (normalized) distances of genomic regions in relation to the radii of cell nuclei (e.g., Cremer et al. 2001; Kozubek et al. 2002). In addition, features like volume, roundness factor, and degree of overlap between genomic regions were used to characterize the geometry of genomic regions (e.g., Götze et al. 2007). Although these techniques were successfully applied, only basic geometric features have been determined and evaluated. Thus, the geometry of the considered objects is not fully exploited and the interpretation of the higher-level geometry is difficult. The reason is that the features (e.g., distances) can be straightforwardly calculated based on the 3D coordinates of the genomic regions, whereas the converse is not generally true Given the features, it is generally not straightforward and often even not possible to uniquely determine the 3D coordinates (e.g., Dryden and Mardia 1998; Adams et al. 2004). In fact, the approaches used in previous studies did not allow integration of the basic geometric features to obtain knowledge about the higher-level chromosomal geometry.

In this work, we *directly* exploit the higher-level geometric structure of the chromatin data. Our work is based on 3D geometrical morphometrics within statistical shape theory (e.g., Rohlf and Marcus 1993; Dryden and Mardia 1998) which we propose to use in conjunction with registration methods for spatial normalization. Whereas in our previous work (Yang et al. 2007) we used image data of *three* consecutive genomic regions (BACs) within one image, we here analyze image data of *four* consecutive BACs and we also use different analysis techniques, e.g., statistical modeling with the Fisher distribution. Our work is based on multi-channel 3D cell microscopy images stained by FISH (fluorescence in situ hybridization). In each image, *four* consecutive genomic regions (BACs) have been labeled on the Xa- and the Xi-chromosome (see Fig. 4.1). The geometric centers of the BAC signals were determined after threshold-based segmentation. In our application we have used two sets of data, where each set represents four consecutive BACs in a number of individual fibroblast cells and three BACs are overlapping, i.e. have been labeled in both datasets. Thus, dataset 1 contains the coordinates of BAC2, BAC3, BAC4, BAC5, and dataset 2 represents the positions of BAC1, BAC2, BAC3, BAC4. Note, that this constellation of BACs in the two datasets allows a sound combination of the geometric information in the two datasets, and to *uniquely* determine the *3D structure* of *five* consecutive BACs. Note also, that in the 2D case already two overlapping consecutive BACs would be sufficient to uniquely determine the *2D structure.* To study the 3D structure of the generated two datasets we used the following approaches. First, a shape uniformity test was used for the 3D structure formed by the four BACs, in order to determine whether the structures are random or not. Second, we separated each tetrahedron (representing four consecutive genomic regions) into four triangles (BAC-triangles), which were then transformed onto the xy-plane using 3D point-based

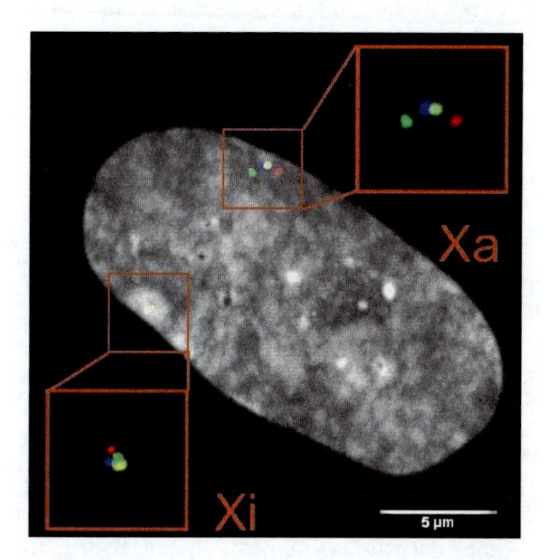

Fig. 4.1 Maximum intensity projection of a 3D confocal microscopy image of a nucleus with four segmented BAC signals in each of the two X-chromosome territories. Active (Xa) and inactive chromosome (Xi)

registration. After using Kendall's spherical coordinate system the shape distribution of BAC-triangles can be considered as a standard problem on a spherical space. Subsequently, the Fisher distribution was applied to statistically model the spherical data. Here, a goodness-of-fit test was used to find out, whether the Fisher distribution well describes the real data. Based on this, a two-sample test was performed to compare the 3D structures of the Xa- and Xi-chromosome. Furthermore, the correlation coefficients on spherical space were also computed to characterize the statistical (in) dependence between the shapes of Xa and Xi. Since both datasets possess an overlap of three BACs, we were able to reconstruct the 3D mean structure formed by *five* consecutive BACs. To this end we applied a point-based rigid registration approach. Moreover, we studied the 3D structure before and after geometric normalization with respect to the cell nucleus shape and analyzed the influence of the normalization on the position of the genomic regions on the basis of a non-rigid image registration approach.

4.2 Image Data and Computational Methods

4.2.1 Image Data

For labeling four consecutive genomic subregions of a human X-chromosome, fluorescence in situ hybridization (FISH) with bacterial artificial chromosome (BAC) probes was used for 3D-preserved nuclei of normal diploid human fibroblasts (Solovei et al. 2002; Cremer et al. 2007). Four consecutive BACs and the inter-BAC regions encompassed about 3.7 Mb in the subcentromere region on the long arm of the X-chromosome. Hybridized probes were detected using different fluorochromes and nuclear DNA was counterstained with DAPI. 3D images (image stacks) were acquired using a Leica SP5 confocal microscope. Stacks of 8-bit gray-scale images were obtained with axial distances of 120 nm between optical sections and pixel sizes of 45 nm. The four consecutive BAC probes, which form a tetrahedron, were visualized in four separate image channels (see Fig. 4.1). Since both X-chromosome homologues (active and inactive X-chromosomes, Xa and Xi) have the same genes, each multichannel image actually contains eight genomic regions, representing two homologous tetrahedrons for the Xa- and Xi-chromosomes. We have segmented the images by global thresholding and the geometric centers of the BACs have been computed. To avoid bias caused by chromatic shift between channels, the coordinates of the signals were corrected for chromatic shift measured as described in Walter et al. (2006). In this study we have used two different datasets: dataset 1 contains the coordinates of BAC2, BAC3, BAC4, BAC5 (in 33 3D multichannel images of different cells) and dataset 2 represents the positions of BAC1, BAC2, BAC3, BAC4 (in 38 different 3D multichannel images). Thus, *three* BACs, namely BAC2, BAC3, BAC4, have been labeled in both datasets. The different BACs are associated with colors as follows: BAC1 (dark blue), BAC2 (green), BAC3 (cyan), BAC4 (yellow), and BAC5 (red).

Note, that the three "overlapping BACs" introduce a common coordinate system which enables spatial alignment of the two datasets. Thus, it is possible to uniquely determine the 3D structure formed by *five* consecutive genomic regions.

Besides the real datasets we also generated two sets of simulated data, which serve as reference datasets. First, a "stable" dataset was created by defining 50 tetrahedrons whose vertices were isotropic normally distributed $N(\mu, \sigma)$, where μ represents the positions of four points and σ was chosen to be small. Second, a "random" dataset including 50 tetrahedrons was created, whose vertices were obtained by random selection within a unit cube. Analogously, we have also created two simulated datasets for the case of three points representing a triangle.

4.2.2 Uniformity Test

In our work, it is essential to first figure out whether the acquired data of the X-chromosome has uniform distribution on the shape space, i.e., whether the shapes of the given data are random or not. We first introduce some basic concepts of geometric morphometrics in statistical shape theory. Given a configuration matrix with k labeled m-dimensional points $\mathbf{X}:(\mathbf{x}_1,\ldots,\mathbf{x}_k)$ in $\mathbb{R}^{m\times k}$ that capture the geometric information of a certain object. The *shape* of the configuration matrix $\mathbf{X}$ is the geometric information, which is invariant under translation, rotation, and isotropic scaling (Euclidean similarity transformation) (Mardia and Jupp 2000). The corresponding shape space Σ_m^k is the set of all possible shapes. To study the shape, it is necessary to first remove the effects of the Euclidean similarity transformation. For example, centering and rescaling $\mathbf{X}$ can produce the *pre-shape* $\mathbf{Z}$ by means of:

$$\mathbf{Z} = [\mathrm{tr}(\mathbf{X}\mathbf{H}^T\mathbf{H}\mathbf{X}^T)]^{-1/2}\,\mathbf{X}\mathbf{H}^T \tag{4.1}$$

where $\mathbf{H}$ represents the $(k-1)\times k$ Helmert submatrix. Then the shape can be represented by $[\mathbf{Z}]$, where $[\mathbf{Z}]=\mathbf{U}\mathbf{Z}, \mathbf{U}\in SO(m)$, and $SO(m)$ represents the orthogonal group of rotations. On the other hand, to remove the rotation different approaches can be used, e.g., point-based registration or generalized Procrustes analysis.

Chikuse and Jupp (2004) have proposed a statistical test to examine the uniformity of a shape as statistical variable. Suppose we are given n samples $\mathbf{X}_1,\ldots,\mathbf{X}_n$ (e.g., the BAC configuration of four consecutive points as in our case), and we have the null-hypothesis that the given data has uniform distribution on the shape space. The statistical test in Chikuse and Jupp (2004) is based on:

$$V = \frac{n(k-1)[m(k-1)+2]\mathrm{tr}(\overline{\mathbf{T}}^2)}{2} \tag{4.2}$$

where $\bar{\mathbf{T}} = \frac{1}{n}\sum_{i=1}^{n}(\mathbf{Z}_i^T \mathbf{Z}_i - \frac{1}{k-1}\mathbf{I}_{k-1})$, and $\mathbf{I}_{k-1}$ is the $k-1$-dimensional identity matrix. Since $(\mathbf{UZ})^T(\mathbf{UZ}) = \mathbf{Z}^T\mathbf{Z}$ holds for any rotation matrix $\mathbf{U}$, the test is rotation invariant. If the shape variables are independent and uniformly distributed, then $V \sim \chi^2_{(k-2)(k+1)/2}$. The null-hypothesis is rejected for large values of V with respect to a specific significance level.

4.2.3 Triangulation

In the section above, we have described a uniformity test for the shape distribution. Assume that the shapes formed by the BACs are not uniformly distributed on the shape space, then one would also like to determine the local *shape* of these BACs. Note that four consecutive BACs form a tetrahedron with four triangular faces (BAC-triangles). The four groups of BAC-triangles were denoted as TR1 to TR4 (see Table 4.1).

4.2.4 Point-Based Rigid Registration

The motivation for using point-based rigid registration in our work is twofold. First, prior to analyzing the BAC-triangles using statistical shape theory, it is necessary to transform each vertex of the BAC-triangles from 3D to 2D coordinates. To this end we suggest employing 3D point-based rigid registration (translation, rotation) to transform all 3D BAC-triangles onto the xy-plane (reference system). The second reason is that we can fix three of the four BACs to analyze the spatial distribution of the remaining "free" BAC, and using this approach we are also able to geometrically combine the two datasets to determine the mean structure of *five* consecutive BACs.

3D point-based rigid registration can be formulated as follows. Given k source points $\mathbf{p}_i$, and k target points $\mathbf{q}_i$, the task is to find a rigid transformation $\mathbf{R}$, such that $\sum_{i=1}^{k}\left\|\mathbf{q}_i - \mathbf{p}_i \circ \mathbf{R}\right\|^2$ is minimized. To transform BAC-triangles ($k=3$) onto the xy-plane, we compute the mean triangle over all images from one dataset. The mean triangle serves as the target points for point-based registration. For point-based registration we employ the quaternion-based algorithm of Horn (Horn 1987). Note, that originally the BAC-triangles are labeled clockwise or counter-clockwise. However, after point-based registration, there is exclusively one kind of labeling order, i.e. either only clockwise or only counter-clockwise, because

Table 4.1 BAC-triangles TR1,...,TR4 and corresponding image channels

Triangles	TR1	TR2	TR3	TR4
Channels	1,2,3	1,3,4	1,2,4	2,3,4

counter-clockwise order and clockwise order can be transformed to each other by a 3D rotation. This is called removing the reflection shape (Dryden and Mardia 1998).

4.2.5 *Investigation of the Statistical Distribution Using Kendall's Spherical Coordinates*

To investigate the statistical distribution on the shape space of triangles we use Kendall's spherical coordinate system. With this coordinate system each triangle is mapped to one point on a sphere. The points on the southern hemisphere represent the reflection shape of those triangles on the northern hemisphere. Furthermore, the two poles of the sphere correspond to an equilateral triangle and its reflection shape, whereas flat triangles are found in the regions close to the equator. In our case, the reflection shapes of the triangles have been removed after 3D point-based registration. Hence, we need to consider only one hemisphere of Kendall's spherical coordinate system.

4.2.6 *Statistical Shape Modeling Using the Fisher Distribution*

After having mapped the triangles to points on a sphere, we suggest modeling the resulting point distribution using a method from spherical statistics. The Fisher distribution in spherical statistics is equivalent to the normal distribution in conventional statistics (Mardia and Jupp 2000). The Fisher distribution is the basic model for spherical data with rotational symmetry, and serves more generally as an all-purpose probability model for spherical data (Fisher et al. 1987). This model allows us to perform a comprehensive statistical inference on the spherical data.

The probability density function of the Fisher distribution is:

$$f(\mathbf{x}) = \frac{\kappa}{\sinh(\kappa)} \exp(\kappa \mu^T \mathbf{x}) \tag{4.3}$$

where $\mathbf{x}$ is a unit vector on a unit sphere, κ is the concentration parameter ($\kappa \geq 0$), and μ denotes the *mean direction*, where $\mu \in \mathbb{R}^3, \|\mu\| = 1$.

Before performing statistical inference, one should analyze how well the Fisher distribution fits the real datasets. The null-hypothesis is that the point distribution of the given spherical data is a Fisher distribution. The formal goodness-of-fit test comprises three subtests (colatitude test, longitude test, and two variable test). For details of this test we refer to Fisher et al. (1987).

Generally, based on the Fisher distribution, we are able to apply different statistical inferences on the spherical data. For example, the two-sample test is used to examine whether Xa and Xi have the same concentration parameter κ of the Fisher distribution.

4.2.7 *Quantitative Analysis of the Correlation Between the Shapes of the Active and Inactive X-Chromosomes*

Each female cell has a pair of X-chromosomes, the active X-chromosome (Xa) and the incative X-chromosome (Xi). An essential question is whether there exists a geometric dependency between the *shapes* of the BAC-triangles of Xa and Xi. This question can be answered using the correlation coefficient between two spherical datasets. In our case, we used the approach of Jupp and Mardia (1980) to assess the dependence between BAC-triangles on Xa- and Xi-chromosomes.

4.2.8 *Spatial Normalization of Cell Nuclei Using Non-rigid Registration*

When studying the 3D structure of chromosomes and genomic regions, a central question is whether the overall cell nucleus shape has an influence on the result. This is particularly relevant in our case, since the analyzed BACs have been labeled in individual cell nuclei which generally have different shapes. If the shape of the cell nuclei has an influence on the location of the BACs, then the question of how to remove potential artifacts due to shape variations becomes an essential issue. One main approach to this problem is to geometrically align and spatially normalize different cell nuclei, which can be achieved using registration techniques. In this work, we used the *non-rigid* intensity-based registration approach of Yang et al. (2008) which allows us to cope with elastic deformations between 3D images. The effect of this approach can be illustrated by Fig. 4.2. Here, 3D images of four nuclei of HeLa cells are shown, where one of the cell nuclei was used as reference (displayed as a dark red object in the background). Within the cell nuclei, chromosome regions (chromosome territories) and genomic regions were labeled. Before registration, some chromosomes territories (indicated by the arrows) lie outside the reference cell nucleus (Fig. 4.2, top middle). After registration, all chromosome territories lie inside the reference cell nucleus and the genomic regions from the different cells are located more closely together (Fig. 4.2, bottom middle and right). In this work, we analyze cell nuclei of human fibroblasts. Since the shape of these nuclei can be approximately described by a 3D ellipsoid, our idea consists in spatially normalizing all cell nuclei with respect to a mean ellipsoid. To determine the mean ellipsoid, i.e. to compute the lengths of the three major axes of the ellipsoid, we use intensity-based *rigid* registration of all cell nuclei. Subsequently, each cell nucleus is registered to this mean ellipsoid using intensity-based non-rigid registration (see Fig. 4.3).

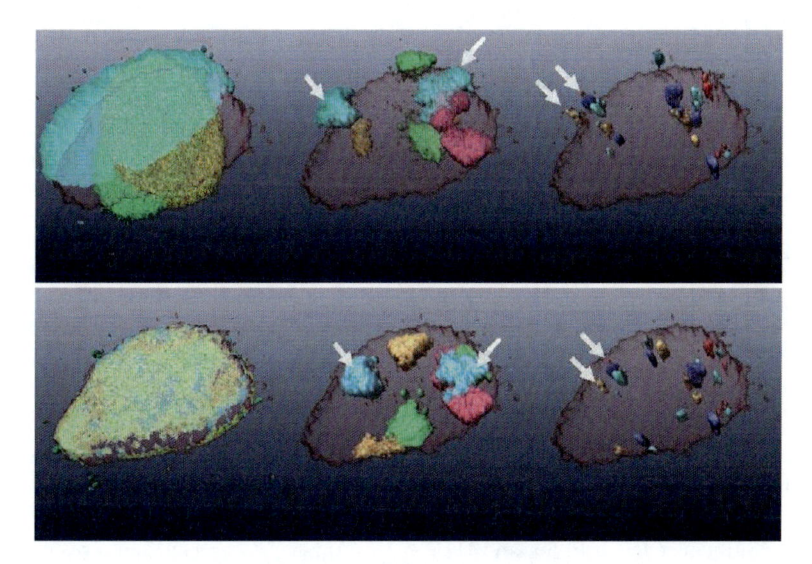

Fig. 4.2 *Top*: 3D overlay of four different cell nucleus images (*left*) with labeled chromosome regions (chromosome territories) (*middle*), and labeled genomic regions (*right*) before registration, *Bottom*: Corresponding images after registration overlaid with a nucleus of one cell as a reference. The *arrows* point to corresponding chromosome territories and genomic regions

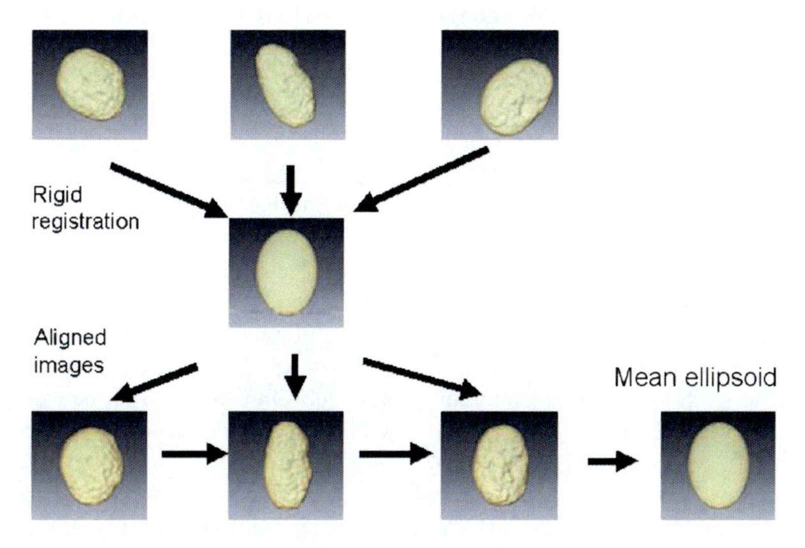

Fig. 4.3 Non-rigid registration of cell nuclei to a mean ellipsoid

4.3 Experimental Results

To analyze the 3D structure of active and inactive X-chromosomes (Xa and Xi) we have applied the uniformity test as described above. The outcome of the uniformity test for four BACs is listed in Table 4.2. The result gives strong evidence that the

Table 4.2 Uniformity test on the shape space Σ_3^4 for a significance level 1% of χ_5^2, threshold value $\chi^2_{5,0.01} = 15.09$ (The threshold value at 5% significance level is $\chi^2_{5,0.05} = 11.07$)

	Dataset 1		Dataset 2	
	Xa	Xi	Xa	Xi
Test value	32.58	38.51	48.10	57.04
Null-hypothesis	R	R	R	R

Null-hypothesis, the shape distribution is uniform; R, null-hypothesis rejected; NR, null-hypothesis not rejected.

shapes of Xa and Xi are non-random. The uniformity test was also used to verify the simulated "stable" dataset as well as the "random" dataset (see Section 4.2.1). The obtained test values are $V_s = 153.09$ and $V_r = 1.46$, respectively. Considering two different significance levels of 5% and 1% ($\chi^2_{5,0.01} = 15.09$, $\chi^2_{5,0.05} = 11.07$), it turns out that the test value for the stable data is indeed much larger than the thresholds (i.e., the data is quite different from random data), whereas the random data exhibits a significantly smaller test value than the thresholds.

As described above in Section 4.2.6, the goodness-of-fit test is used to examine how well the Fisher distribution fits our real data. The results of this test shows that all three test values of all BAC-triangles of dataset 1 are smaller than the threshold values for the significance level of 1%, i.e., $M^c_{0.01} = 1.308$, $M^l_{0.01} = 1.347$, $M^t_{0.01} = 1.035$. Therefore, the shape distribution of all BAC-triangles in dataset 1 follows a Fisher distribution for a significance level of 1%. However, for dataset 2 most of the measurements can be also modeled by a Fisher distribution except for the BAC-triangle configuration TR3 of Xa and TR2 of Xi.

On the basis of the goodness-of-fit tests, we further investigated the parameters of the Fisher distribution for the two real datasets (see Table 4.3). In our previous work (Yang et al. 2007), we have used the complex Bingham distribution to statistically describe the shape distribution of BAC-triangles. Here, in order to be able to compare the results we additionally determined the corresponding concentration parameters λ of the complex Bingham distribution (see Table 4.3). Also, the estimated parameters for the simulated data of the triangles (see Section 4.2.1) are provided as a reference. For the random dataset we obtained $\hat{\mu}_r = (-0.04, 0.08, 0.99)$, $\hat{\kappa}_r = 2.65$, $|\hat{\lambda}_r| = 5.63$, and for the stable dataset we yielded $\hat{\mu}_s = (0.34, 0.04, 0.94)$, $\hat{\kappa}_s = 81.27$, $|\hat{\lambda}_s| = 162.53$. Comparing these values with the results for the real data, it turns out that the real data exhibits a relatively large variation in shape, since κ and λ are significantly smaller. Finally, we need to answer the important question, whether the BAC-triangles have uniform distribution on the shape space. The shape uniformity test for BAC-triangles was performed analogously to the approach in Yang et al. (2007). However, in contrast to that work, we here used a more robust uniformity test i.e. Giné's F_n test (Mardia and Jupp 2000). Analyzing the test results, we found that the *local* BAC-triangles are highly variable and some of them even resemble a random shape, although all *global* shapes of the BAC-tetrahedrons had been proven to be not random.

Table 4.3 Parameters of the Fisher distribution for dataset 1. $\hat{\mu}_x, \hat{\mu}_y, \hat{\mu}_z$ are the three components of the *mean direction* $\hat{\mu}$, where $\| \hat{\mu} \|=1$. $\hat{\kappa}$ denotes the estimated concentration parameter of the Fisher distribution, whereas $|\hat{\lambda}|$ represents the absolute value of the estimated concentration parameter of the complex Bingham distribution. The uniformity test for the BAC-triangles is Giné's F_n test with a significance level of 1%, $F_{n,0.01}=3.633$ (The threshold for a significance level of 5% is $F_{n,0.05}=2.748$)

	Dataset 1							
	Xa				Xi			
	TR1	TR2	TR3	TR4	TR1	TR2	TR3	TR4
$\hat{\mu}_x$	0.79	0.51	0.27	–0.09	–0.06	0.38	0.04	0.17
$\hat{\mu}_y$	0.07	–0.20	–0.51	0.38	0.08	–0.27	–0.68	0.44
$\hat{\mu}_z$	0.61	0.83	0.82	0.92	0.99	0.88	0.73	0.88
$\hat{\kappa}$	3.20	2.87	3.78	3.53	2.50	2.88	3.27	3.18
$\|\hat{\lambda}\|$	7.84	6.26	9.06	9.54	5.24	6.25	7.47	7.17
Test value	3.12	3.67	7.26	5.05	1.16	4.24	7.46	4.19
Null-hypothesis	NR	R	R	R	NR	R	R	R

Null-hypothesis: The shape distribution is uniform. R: Null-hypothesis rejected, NR: Null-hypothesis not rejected.

The two-sample test determines whether both samples of Xa and Xi have the same concentration parameter in terms of the Fisher distribution. Given a significance level of 1%, the results show that all BAC-triangles of Xa and Xi in dataset 1 have the same concentration parameters κ, while not all BAC-triangles of dataset 2 exhibit this property.

Subsequently, we evaluated the dependence between Xa and Xi using the spherical correlation coefficient. Given a significance level of 1%, Table 4.4 shows that the shapes of all BAC triangles of Xa and Xi are independent of each other. However, TR3 of Xa and Xi in dataset 2 possesses a dependence for a significance level of 5%, which is interesting to point out.

As mentioned in Section 4.2.8 above, we have studied the effects of geometric normalization of cell nuclei using non-rigid image registration. Table 4.5 shows that the *V*-value of the uniformity test (see Section 4.2.2) decreased significantly after normalization (except for Xa of dataset 2). The results indicate that the structures after normalization exhibit a higher variation and that the effect is not negligible.

We have also prepared 3D visualizations of the result after 3D point-based rigid registration (see Section 4.2.4). The visualization allows a qualitative assessment concerning the randomness of the investigated real datasets. The registration yielded different clusters as shown in Fig. 4.4, where the red (BAC5) (see Figs. 4.4a,b) and the dark blue (BAC1) (see Figs. 4.4c,d) cluster are the "free" BACs with respect to the reference system established by the other three fixed BACs.

Table 4.4 Correlation coefficient (CC) and independence test between Xa and Xi for a significant level of 1%, the corresponding threshold value is $\chi^2_{9,0.01} = 21.66$. Hence, there exists no evidence for a dependency. However, if the significance level is set to 5%, i.e. $\chi^2_{9,0.05} = 16.92$, then Xa and Xi of TR3 of dataset 2 exhibit a weak dependency

	Dataset 1				Dataset 2			
	TR1	TR2	TR3	TR4	TR1	TR2	TR3	TR4
CC	0.15	0.31	0.14	0.32	0.39	0.34	0.50	0.18
Test value	4.89	10.12	4.60	10.45	14.79	12.99	19.18	7.04
Null-hypothesis	NR	NR	NR	NR	NR	NR	NR	NR

Null-hypothesis: Both datasets are independent of each other. R: Null-hypothesis rejected, NR: Null-hypothesis not rejected.

Table 4.5 Uniformity test for four BACs (see Section 4.2.2) before and after normalization by non-rigid registration

	Dataset 1		Dataset 2	
	Xa	Xi	Xa	Xi
V before normalization	32.58	38.51	48.10	56.97
V after normalization	21.31	25.98	46.02	44.17

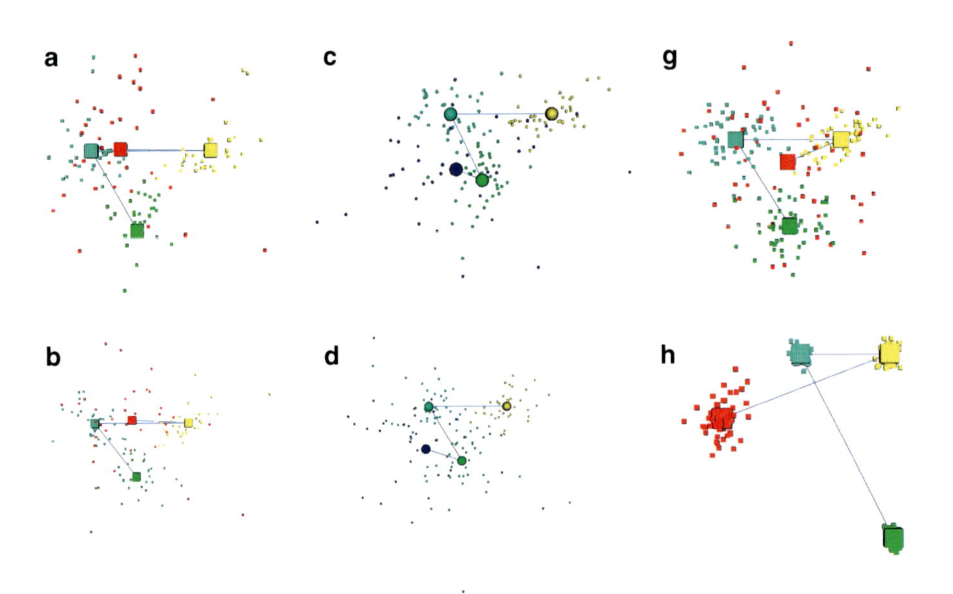

Fig. 4.4 Results of point-based registration using three overlapping BACs, z-projections (BAC1, dark blue; BAC2, green; BAC3, cyan; BAC4, yellow; BAC5, red) (**a**) Xa of dataset 1; (**b**) Xi of dataset 1; (**c**) Xa of dataset 2; (**d**) Xi of dataset 2; (**g**) result for random dataset; (**h**) result for stable dataset. Note that for dataset 1 the genomic sequence of the BACs is visualized by the colors green, cyan, yellow, and red. For dataset 2 the genomic sequence is dark blue, green, cyan, and yellow

The larger cubes and spheres visualize the mean positions of the clusters for dataset 1 and 2, respectively. Furthermore, we have applied the same registration approach to both simulated datasets. Considering the result for the stable dataset (see Fig. 4.4h), we found that the structures for the real data differ significantly. On the other hand, for the simulated random dataset we obtained always the same mean structure as displayed in Fig. 4.4g, even though performing the simulation repeatedly. We found that the mean positions of the real data differ also remarkably from that of the random data, which indicates also that the real data does not possess a random structure.

In addition , we have reconstructed the 3D structure of *five* consecutive BACs by combining the 3D structure from four overlapping consecutive BACs. Figure 4.5 shows the results after point-based registration (see Section 4.2.4). Here, the mean positions of the green, cyan, and yellow points serve as reference points building a common coordinate system. Based on this coordinate system, dataset 2 was aligned to dataset 1, and we were therefore able to study both positions of the red and dark blue points in relation to the other three reference points. In Fig. 4.5 (top left) the two datasets were overlaid with each other. Finally, in order to obtain a mean structure, the two registered datasets were merged (see Fig. 4.5 bottom). The result for Xi is similar to that of Xa, therefore a visualization has not been displayed. Finally,

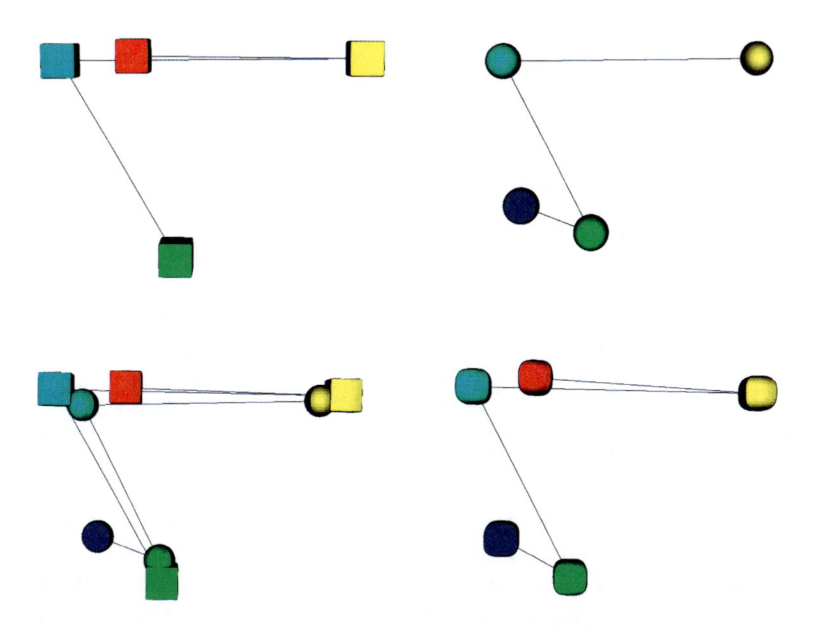

Fig. 4.5 3D Reconstruction of mean positions of *five* BACs by fixing three of them (here for Xa). *Top left*: Mean positions of four BACs of dataset 1. *Top right*: Mean positions of four BACs of dataset 2. *Bottom left*: Result after point-based registration by fixing 3 BACs (cyan, yellow and green). *Bottom right*: Same as bottom left but both datasets are merged. Note, that only mean positions have been displayed. For dataset 1 the genomic sequence of the BACs is green, cyan, yellow, and red. For dataset 2 the genomic sequence of the BACs is dark blue, green, cyan, and yellow

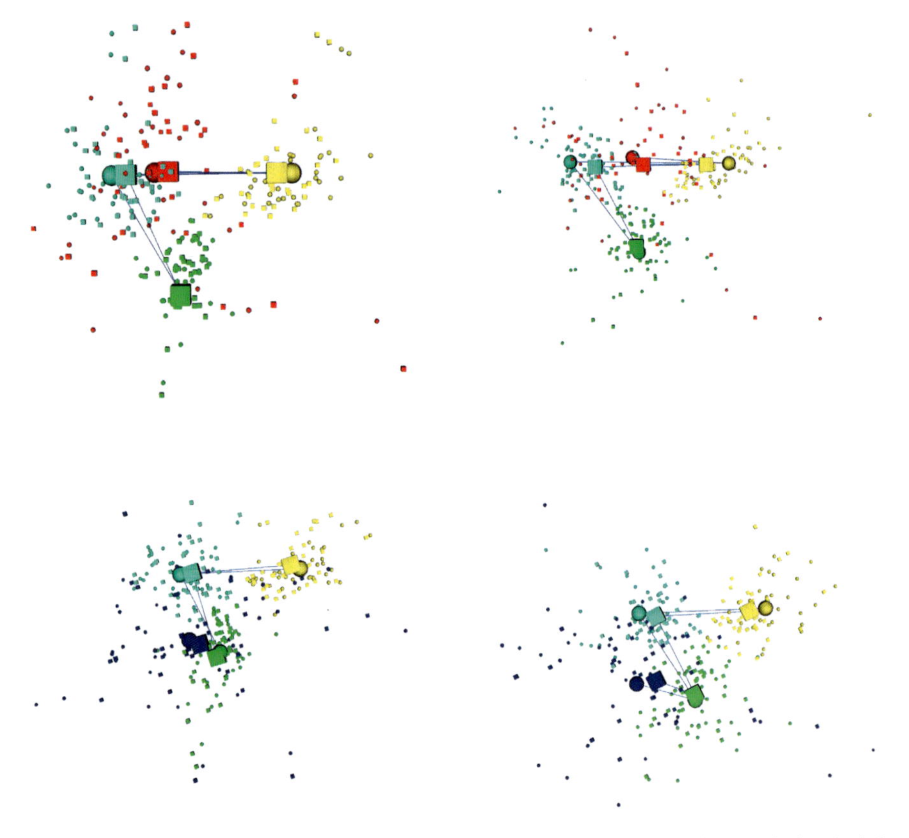

Fig. 4.6 *Top left*: Superposition of the four Xa BACs of dataset 1 before (*sphere*) and after (*cube*) normalization; *Top right*: Same as top left but for Xi BACs of dataset 1; *Bottom left*: Same as top left but for Xa BACs of dataset 2; *Bottom right*: Same as top left but for Xi BACs of dataset 2

to visualize the effect of spatial normalization of cell nuclei we overlaid the mean position of BACs before and after normalization (see Fig. 4.6).

4.4 Discussion

We have systematically investigated the 3D structure formed by four consecutive overlapping BACs on the interphase X-chromosome. Our analysis is based on geometrical morphometrics from statistical shape theory in conjunction with rigid as well as non-rigid registration methods. In contrast to previous work, where the 3D chromatin structure was analyzed *indirectly* (e.g., based on distances between genomic regions) we here *directly* exploit the 3D coordinates of genomic regions to determine the higher-level structure of chromatin fiber. To analyze the shapes formed by four consecutive genomic regions we have used a uniformity test. Then, a series of evaluations based on the Fisher distribution have been performed to study the shape of BAC-triangles formed by each three of the four BACs. Finally, we have analyzed the structure of five consecutive BACs resulting from two sets of four consecutive BACs.

Our findings can be summarized as follows. First, the goodness-of-fit test for the Fisher distribution provides evidence that most of the BAC-triangles follow a Fisher distribution. Second, the estimated concentration parameters of this distribution model show that the shape variations of the BAC-triangles are relatively high. Third, using the two-sample test, we have examined the similarity between the BAC-triangles of Xa- and Xi-chromosomes in terms of concentration parameters. It turned out that the first dataset has similar concentration parameters for Xa and Xi, whereas the second dataset exhibits some differences. Fourth, we have evaluated the spherical correlation coefficient of the BAC-triangles for Xa and Xi to assess their statistical dependence. It turned out, that the BAC-triangles of Xi and Xa are statistically independent except for one case in the second dataset. Moreover, we have used point-based registration to reconstruct and visualize the average spatial 3D shape of five consecutive BACs. With regard to the simulated random data, the visualization provides qualitative evidence that the shapes of the investigated data are not random. Finally, we have used non-rigid image registration to spatially normalize the shape of cell nuclei with respect to a mean ellipsoid (representing the mean shape of the cell nuclei). Comparing the results before and after normalization, it turned out that we obtain a significant effect with respect to the location of genomic regions which suggests that such type of shape normalization should be generally used for accurately analyzing nuclear organization.

In conclusion, statistical shape theory in conjunction with registration methods as proposed here provides a sound framework to support research on the 3D structure of chromosomes. The described approaches can also be applied in related studies on nuclear architecture .

Acknowledgements This work has been supported by the EU project 3DGENOME. TC also acknowledges the support by the Wilhelm-Sanderstiftung (2001.079.2). The work benefited from the use of the Insight Toolkit (ITK) Ibanez et al. (2005) and the Visualization Toolkit (VTK) (Schroeder et al. 2003).

References

Adams DC, Rohlf FJ, Slice DE (2004) Geometric morphometrics: ten years of progress following the 'revolution'. Ital J Zool 71:5–16

Chikuse Y, Jupp PE (2004) A test of uniformity on shape spaces. J Multivar Anal 88:163–176

Chuang C, Belmont A (2007) Moving chromatin within the interphase nucleus-controlled transitions? Semin Cell Dev Biol 18:698–706

Chuang C, Carpenter A, Fuchsova B, Johnson T, de Lanerole P, Belmont AS (2006) Long-range directional movement of an interphase chromosome site. Curr Biol 16:825–831

Cremer M, von Hase J, Volm T, Brero A, Kreth G, Walter J, Fischer C, Solovei I, Cremer C, Cremer T (2001) Non-random radial higher-order chromatin arrangements in nuclei of diploid human cells. Chromo Res 9(7):541–567

Cremer M, Müller S, Köhler D, Brero A, Solovei I (2007) Cell preparation and multicolor FISH in 3D preserved cultured mammalian cells, CSH Protocols, doi:10.1101/pdb.prot4723

Dryden I, Mardia K (1998) Statistical Shape Analysis. Wiley, Chichester

Dundr M, Ospina J, Sung M, John S, Upender M, Ried T, Hager G, Matera A (2007) Actin-dependent intranuclear repositioning of an active gene locus in vivo. J Cell Biol 179:1095–1103

Fisher NI, Lewis T, Embleton B (1987) Statistical analysis of spherical data. The Press Syndicate of the University of Cambridge, Cambridge

Fraser P, Bickmore W (2007) Nuclear organization of the genome and the potential for gene regulation. Nature 447:413–417

Götze S, Mateos-Langerak J, Gierman HJ, de Leeuw W, Giromus O, Indemans MHG, Koster J, Ondrej V, Versteeg R, van Driel R (2007) The three-dimensional structure of human interphase chromosomes is related to the transcriptome map. Mol Cell Biol 27(12):4475–4487

Horn B (1987) Closed-form solution of absolute orientation using unit quaternions. J Opt Soc Am A 4:629–642

Ibanez L, Schroeder W, Ng L, Cates J (2005) The ITK software guide. Kitware, New York

Jupp PE, Mardia KV (1980) A general correlation coefficient for directional data and related regression problems. Biometrika 67(1):163–173

Kosak S, Groudine M (2004) Form follows function: the genomic organization of cellular differentiation. Genes Dev 18:1371–1384

Kozubek S, Lukásová E, Jirsová P, Koutná I, Kozubek M, Ganová A, Bártová E, Falk M, Paseková R (2002) 3D structure of the human genome: order in randomness. Chromosoma 111:321–331

Lanctôt C, Cheutin T, Cremer M, Cavalli G, Cremer T (2007) Dynamic genome architecture in the nuclear space: regulation of gene expression in three dimensions. Nat Rev Genet 8(2):104–115

Lyon M (1961) Gene action in the x-chromosome of the mouse (mus musculus l.). Nature 190:372–373

Mardia KV, Jupp PE (2000) Directional statistics. Wiley, Chichester

Ostashevsky J (1998) A polymer model for the structural organization of chromatin loops and minibands in interphase chromosomes. Mol Biol Cell 9:3031–3040

Parreira L, Telhada M, Ramos C, Hernandez R, Neves H, Carmo-Fonseca M (1997) The spatial distribution of human immunoglobulin genes within the nucleus: evidence for gene topography independent of cell type and transcriptional activity. Hum Genet 100:588–594

Ragoczy T, Bender M, Telling A, Byron R, Groudine M (2006) The locus control region is required for association of the murine beta-globin locus with engaged transcription factories during erythroid maturation. Genes Dev 20:1447–1457

Rohlf FJ, Marcus IF (1993) A revolution in morphometrics. Trends Ecol Evol 8:129–132

Schroeder W, Matin K, Lorensen B (2003) The visualization toolkit, an object-oriented approach to 3D graphics, 3rd edn. Kitware, New York

Sexton T, Schober H, Fraser P, Gasser S (2007) Gene regulation through nuclear organization. Nat Struct Mol Biol 14:1049–1055

Skalníková M, Kozubek S, Lukášova E, Bártová E, Jirsová P, Cafourková A, Koutná I, Kozubek M (2000) Spatial arrangement of genes, centromeres and chromosomes in human blood cell nuclei and its changes during the cell cycle, differentiation and after irradiation. Chromo Res 8:487–499

Solovei I, Walter J, Cremer M, Habermann F, Schermelleh L, Cremer T (2002) FISH: a practical approach, Ch. 7 FISH on three-dimensionally preserved nuclei. Oxford University Press, Oxford, pp 119–157

Soutoglou E, Misteli T (2007) Mobility and immobility of chromatin in transcription and genome stability. Curr Opin Genet Dev 17:435–442

Sproul D, Gilbert N, Bickmore W (2005) The role of chromatin structure in regulating the expression of clustered genes. Nat Rev Genet 6:775–781

Walter J, Joffe B, Bolzer A, Albiez H, Benedetti P, Müller S, Speicher M, Cremer T, Cremer M, Solovei I (2006) Towards many colors in FISH on 3D-preserved interphase nuclei. Cytogenet Genome Res 114:367–378

Yang S, Götze S, Mateos-Langerak J, van Driel R, Eils R, Rohr K (2007) Variability analysis of the large-scale structure of interphase chromatin fiber based on statistical shape theory. In: Advances in mass-data, analysis of images and signals in medicine, biotechnology and chemistry (MDA). Springer-Verlag, Berlin, pp 37–46

Yang S, Köhler D, Teller K, Cremer T, Baccon PL, Heard E, Eils R, Rohr K (2008) Nonrigid registration of 3-d multichannel microscopy images of cell nuclei. IEEE Trans Image Process 17(4):493–499

Yokota H, van den Engh G, Jearst J, Sachs R, Trask B (1995) Evidence for the organization of chromatin in megabase pair-sized loops arranged along a random walk path in the human g0/g1 interphase nucleus. J Cell Biol 130:1239–1249

Zlatanova J, Leuba S, van Holde K (1998) Chromatin fiber structure: morphology, molecular determinants, structural transitions. Biophys J 74:2554–2566

Chapter 5
Nuclear Molecular Motors for Active, Directed Chromatin Movement in Interphase Nuclei

Joanna M. Bridger and Ishita S. Mehta

Abstract The nucleus is a highly organised organelle that contains structural elements which interact and control the genome. A few studies have started to undercover a role for actin and myosin isoforms, found in the nucleus, as nuclear motors that actively move individual gene loci, clusters of genes and whole chromosomes within the nucleoplasm. This chapter reviews these few studies, discusses the presence of proteins potentially part of acto-myosin nuclear motors and asks where these studies should aim to head in the future.

Keywords Myosin • Actin • G-actin • F-actin • β-actin • Myosin 1 • Nuclear myosin 1β • MYO1C • Nuclear motors • Molecular motors • Actin-related proteins • Arps • Gene re-positioning • Chromosome territories • Emerin • Lamin A • Arp6 • Cajal body • Mlp1

5.1 Molecular Motors

Although, the cell nucleus contains structural elements that bind and anchor other structures there is evidence and need for of rapid movement of entities within the nucleoplasm. Indeed, genes, clusters of genes and even whole chromosomes relocate within nuclei and can travel long distances in quite small amounts of time. How this rapid movement of chromatin is elicited is still being discussed but in the last few

J.M. Bridger (✉)
Laboratory of Nuclear and Genomic Health, Centre for Cell and Chromosome Biology, School of Health Sciences and Social Care, Brunel University, Uxbridge UB8 3PH, London
e-mail: Joanna.bridger@brunel.ac.uk

I.S. Mehta
Laboratory of Nuclear and Genomic Health, Centre for Cell and Chromosome Biology, School of Health Sciences and Social Care, Brunel University, Uxbridge UB8 3PH, London
and
DNA Repair, Recombination and Replication Laboratory, Department of Biological Sciences, Tata Institute of Fundamental Research, Homi Bhabha Road, 400 005 Colaba, Mumbai, India

N.M. Adams and P.S. Freemont (eds.), *Advances in Nuclear Architecture*,
DOI 10.1007/978-90-481-9899-3_5,

years there is a building body of evidence pointing towards a molecular motor activity that actively engages with chromatin and translocates it across nuclei.

These molecular motors may well be similar to those found in the cell cytoplasm. The cytoplasmic motors are well-researched, documented and are stable curricula in cell biology text books and teaching. Molecular motors can be built from a wide range of myosin isoforms (see Myosins, a superfamily of molecular motors edited by Lynne M. Coluccio for Springer, Dordrecht; for an extensive overview of myosin nomenclature, structure and function). Myosins are actin-activated motor molecules that use energy from ATP hydrolysis to generate force and move along actin filaments (Pollard et al. 1973; Pollard and Korn 1973) to instigate various cytoplasmic processes such as cell crawling, cytokinesis, phagocytosis, growth cone extension, maintenance of cell shape, organelle/particle trafficking (Berg et al. 2001), signal transduction (Bahler 2000), establishment of polarity (Yin et al. 2000) and actin polymerisation (Evangelista et al. 2000; Lechler et al. 2000; Lee et al. 2000).

The myosins interact with actin in various forms of polymerisation to perform motor activities for a wide variety of active processes. Molecular motors often work in concert with microtubules or microfilaments to act as work-horses in the cell creating movement – of entities within cells, the cell itself and whole tissues such as in skeletal muscle. Force is required to perform movement within the cell nucleus. The movement of myosin over actin filaments generates force within the cytoplasm which facilitates muscle contraction (Sellers 2000). Thus, with the presence of both actin and myosin isoforms within the cell nucleus, it is highly likely that these two proteins interact in the intra-nuclear environment and thus may be involved in movement of various nuclear entities around the nucleoplasm. There is however, very little evidence of interaction between these two proteins within nuclei and this is partly because the nuclear isoform of myosin 1C is an unconventional myosin (Fig. 5.1) that does not form filaments (Mermall et al. 1998) as well as the filamentous forms of nuclear actin being different as compared to the cytoplasmic actin filaments (McDonald et al. 2006) and hard to visualise. It has been proposed by different scientists over the years that these two proteins will act in

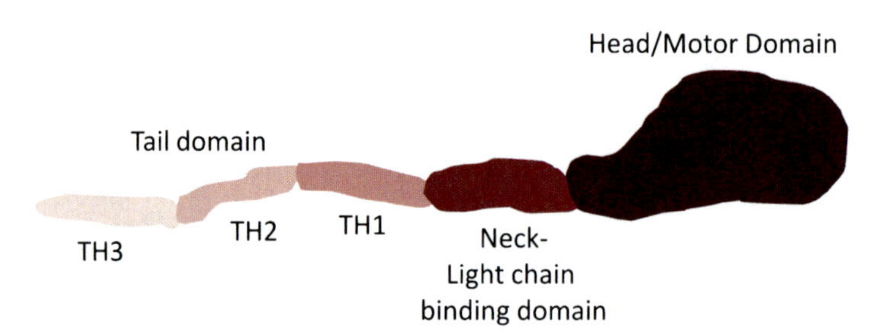

Fig. 5.1 Basic protein structure of myosin 1 in cartoon form. Myosin 1 contains only one globular head which contains a nucleotide binding site and an actin binding site. The next region is the neck or the light-chain-binding domain which is alpha-helical. The tail region is at the C-terminus and has three regions, termed TH1, TH2 and TH3. The function of TH1 is to elicit phospholipid or membrane binding; TH2 contains another actin binding site and TH3 has a domain that is involved in signal transduction, actin dynamics and membrane trafficking

concert to form what are called as nuclear motors (de Lanerolle et al. 2005; Hofmann et al. 2006a). Like cytoplasmic motors that facilitate muscle contraction, these nuclear motors may be involved in movement of various nuclear structures throughout the nucleoplasm (Hofmann et al. 2006a). It is as yet unclear whether these proteins are part of the nuclear scaffolding i.e. the nuclear matrix/nucleoskeleton or just interact with it.

There are a lot of articles entering into the literature presently identifying motor proteins as being involved in many nuclear processes – we will cover these in less detail since they are discussed at length in a number of modern and extremely good reviews (Castano et al. 2010; Dion et al. 2010). However, we will attempt to cover a role for nuclear motors in the nucleus in long range chromosomal and sub-chromosomal movement in interphase nuclei.

5.2 The Dynamic Nucleus

The coordination of nuclear processes such as regulation of gene transcription, chromatin remodelling, genome re-organisation, processing of RNA and DNA replication and repair is regulated in a well organised and functionally structured nucleus. For many of these active processes basic research is uncovering the functional necessity for some kind of motor protein and so these are exciting times for nuclear biologists in revealing yet another level of dynamic architecture in molecular motors that is involved in the orchestration of genome behaviour. For example, genomic movements have been observed whereby clusters of genes come away from whole chromosome territories on loops (Branco and Pombo 2006; Osborne et al. 2007) associating with other nuclear structures such as transcription factories and nuclear bodies (Dundr et al. 2007); larger chromosomal regions have been seen to move long distances across a nucleus (Chuang et al. 2006) and even whole individual chromosomes (Mehta 2005; Mehta et al. 2008; Hu et al. 2008; Mehta et al. 2010) These movements are directed and rapid and do not give the impression that they are either passive or random in their nuclear destination. This conclusion leads to one hypothesis – that there is motor machinery within the nucleus that is energy-driven and targets repositioning of nuclear entities and leads us and many others before us to ask the question "Are there motors within the nucleoplasm, that work in similar ways to the motors that operate in the cytoplasm." It is a very obvious leap to make that a cell that has evolved a method, via specific proteins, for moving itself and structures around in the cytoplasm to use a similar method in the cell nucleus. But even the presence of motor proteins in the nucleus has been notoriously difficult to prove, which makes it even harder to experimentally determine the functional mechanism of nuclear motor activity and the involvement of such proteins in different nuclear processes. The types of protein that would be involved in nuclear motors are the myosins and isoforms of actin. This chapter will review the evidence for motor presence and activity in cell nuclei and discuss the different methodologies and technologies that have and can be used to elucidate further this elusive functional mechanism.

5.3 Acto: Myosin Molecular Motors in Nuclei

Over the years many sceptics have disregarded the presence as cytoplasmic contamination either in the biochemical preparations or by cytoplasmic presence in the nucleus via channels and invaginations of the nuclear envelope. With the development of better antibodies against nuclear specific isoforms of myosin and actin, better imaging facilities and system biology approaches to analysing the genome it is now less hard to defend the presence of nuclear motor proteins.

5.3.1 Nuclear Actin and Actin-Related Proteins

The presence of actin in the nucleus was first suggested as early as 1969 when Lane first observed fibrillar bundles in oocytes treated with actinomycin (Lane 1969). This study was followed by many other studies where actin filaments were isolated from nuclei of different cell types from different organisms (Clark and Rosenbaum 1979; Nakayasu and Ueda 1983; Jockusch et al. 1974). Studies by Clark et al. then reaffirmed the presence of actin in the nucleus by careful hand isolation of nuclei from *Xenopus* oocytes so as to avoid any contamination with cytoplasmic proteins (Clark and Rosenbaum 1979; Clark and Merriam 1977). Early studies from the Rosenbaum and Ueda groups also demonstrated the functional relevance of nuclear actin in maintenance of the nuclear envelope via its attachment to the nuclear matrix (de Lanerolle et al. 2005; Clark and Rosenbaum 1979; Nakayasu and Ueda 1983). Following this, a study performed on HeLa cells suggested that nuclear actin may act as a transcription factor for RNA polymerase B and thus may be involved in the process of transcription (Egly et al. 1984). Direct evidence that proved this hypothesis was provided when inhibition of chromosome condensation (Rungger et al. 1979) and transcription (Scheer et al. 1984) was observed following microinjection of anti-actin antibodies into the nucleus. However, detection of nuclear actin in the nucleus was not well accepted and was thought to be just a contamination of cytoplasmic actin (Ankenbauer et al. 1989). These reasons were enough to disregard all the early work on the presence of actin in the nucleus as just a result of unclean preparations. Yet, all efforts to convince the scientific community of the presence of nuclear actin failed as any attempts to detect actin in the nucleus by fluorescence staining using anti-actin antibodies or actin binding compounds failed. Also, early attempts to detect the most common actin binding partner, myosin in the nucleus were also not fruitful (de Lanerolle et al. 2005).

This point of view regarding the presence of actin has changed radically in last few years with many more studies linking actin to various nuclear functions. The first study that gave convincing evidence of actin's involvement in nuclear processes showed the association of β-actin with chromatin modeling complexes like BAF in activated T-lymphocytes (Zhao et al. 1998). The first evidence of other forms of actin present in the nucleus came from a FRAP assay revealing that actin is present as a dynamic mixture of monomeric, oligomeric and polymeric fibres in the cell nucleus (McDonald et al. 2006). Moreover, the actin polymers observed in the nucleus were very different to those found in the cytoplasm and are also much more dynamic in nature (McDonald et al. 2006). The dynamic nature of the actin in the nucleus was further supported by fluorescence staining of actin in the nucleus whereby studies in two labs using anti-actin antibodies, i.e. 2G2 and 1C7 displayed different patterns of actin staining in the nucleus and suggesting the presence of both G- and F-actin within nuclei of cells (Gonsior et al. 1999; Schoenenberger et al. 2005). A few studies have indicated that actin is part of the nuclear matrix/ nucleoskeleton of cells (Wang et al. 2006; Cruz et al. 2009). However, none of these studies have revealed an actin network such that was seen in *Xenopus* oocytes (Fig. 5.2) (Clark and Rosenbaum 1979; Hofmann et al. 2001; Kiseleva et al. 2004) and was very prominent when F-actin export was not blocked (Bohnsack et al. 2006). The network is sponge-like and throughout the entire nucleoplasm and is critical for nuclear stability. It is as yet not clear if actin networks are present in somatic nuclei but there are certainly discussion about different types of actin isoforms in the nucleus (Jockusch et al. 2006) and not just G (monomeric) and F actin (filamentous). Indeed β-actin is also a candidate for dynamic activities in the nucleus, since employing a β-actin dominant negative mutant in cells specific

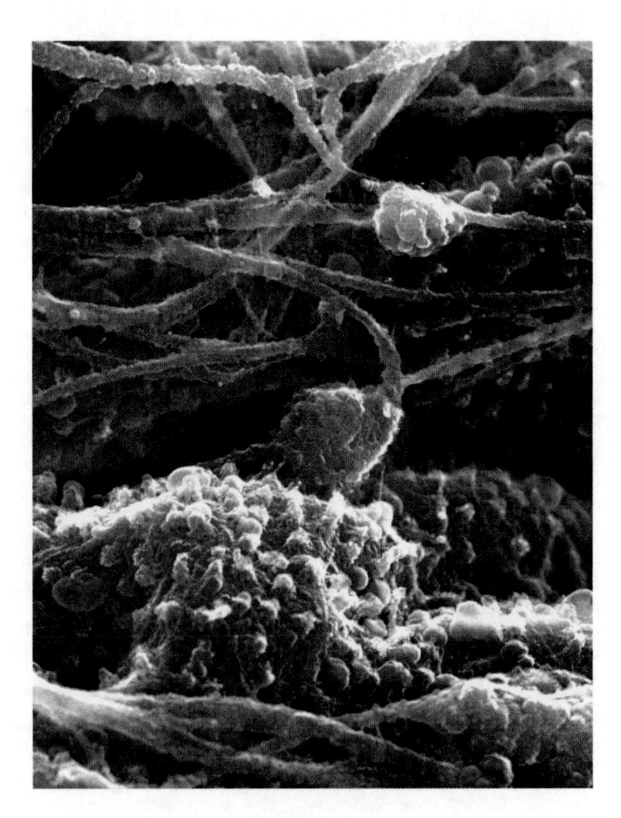

Fig. 5.2 Actin filaments are revealed in a *Xenopus* oocyte using scanning electron microscopy. They are delineated in pink and are shown subjacent to the inner nuclear envelope in grey. Cajal bodies in blue can be observed sitting on the actin filaments. This image was kindly provided by Dr. Elena Kiseleva, Ruissian Academy of Sciences

genes did not translocate to Cajal bodies (Dundr et al. 2007). T-actin is a twisted form of actin filament that may well be present in nuclei and not detectable with phalloidin (Castano et al. 2010; Egelman 2003). Indeed Cajal bodies themselves are co-localised with actin (Gedge et al. 2005) and an acto-myosin activity is required to move PML around nuclei (Muratani et al. 2002).

In this century, there was a sudden surge of reports implicating actin in various nuclear functions including chromatin remodelling (Olave et al. 2002; Pederson 2000), transcription by RNA polymerases (Zhang et al. 2002; Hofmann et al. 2004; Philimonenko et al. 2004; Hu et al. 2004; Kukalev et al. 2005), nuclear transport of RNA (Hofmann et al. 2001; Percipalle et al. 2002; Percipalle et al. 2001; Kimura et al. 2000), controlling the entry of the cells into mitosis (Lee and Song 2007), partitioning of active nuclear components within the nucleoplasm (Andrin and Hendzel 2004) and chromosome decondensation (Gieni and Hendzel 2008). In addition to this, the actin cytoskeleton is also thought to provide a physical bridge between the cytoplasm and the nucleus via its interactions with various INM and ONM proteins such as nuclear lamins and nesprins; and thus might be involved in transmitting mechanical signals from the extracellular matrix to the nucleus (Gieni and Hendzel 2009). Further a protein of the inner nuclear envelope emerin have been shown to have direct interactions with F-actin (Holaska et al. 2004) and nuclear myosin 1β and they are both in a complex that in addition contains A- and B-type lamins and SUN2, which are proteins involved in nuclear architecture (Holaska and Wilson 2007).

In addition to actin there are a plethora of actin-related proteins (Arps) in nuclei that are involved in various remodelling and alteration of chromatin structure but there is now discussion that specific Arps are also involved with chromosome territory structure (Lee et al. 2007) and in repositioning of active chromatin loops to nuclear pore complexes in budding yeast via a myosin-like protein (Dion et al. 2010). Arps are similar to actin with respect to sequence and they display an ATP-binding pocket similar to actin (Castano et al. 2010). Since it also known that actin has a binding affinity for emerin (Holaska and Wilson 2007) and lamin A (Holaska and Wilson 2007; Sasseville and Langelier 1998) it is possible to imagine an actin network functionally interacting with the nuclear periphery (Holaska and Wilson 2007) or deep within the nucleoplasm if lamin A and even emerin are part of a nucleoskeleton (Mehta et al. 2008). Indeed, Hozak and colleagues reveal by immuno-EM that actin is found throughout the nucleoplasm but in localised regions (Dingova et al. 2009).

5.3.2 Nuclear Myosins

Although first identified as cytoplasmic proteins, the presence of myosin isoforms in nuclei of cells has also been discussed over the last 25 years. In 1986, Hagen et al. demonstrated the presence of a nuclear protein that cross-reacted with a monoclonal antibody to *Acanthaamoeba castellanii* myosin 1 (Hagen et al. 1986). Following this study, a myosin II heavy chain like protein was also found near nuclear pore complexes inside the nucleus in *Drosophila melanogaster* (Berrios et al. 1991). Nevertheless, this early work demonstrating the presence of myosin isoforms in the nucleus, like nuclear actin, was met by scepticism and doubt owing to a belief that presence of myosin in the nucleus was also due to contamination by cytoplasmic myosins (Pederson and Aebi 2002; Nowak et al. 1997). Technical advances leading to the identification of various different isoforms of myosin in the nucleus in recent years has changed this point of view dramatically. Two studies from de Lanerolle's group have revealed the presence of nuclear isoform of a myosin 1 (nuclear myosin 1β; NMIβ) in the nucleus which was very similar to myosin 1C found in the cytoplasm (Nowak et al. 1997; Pestic-Dragovich et al. 2000) and is conserved in vertebrates (Kahle et al. 2007). Myosin VI (Vreugde et al. 2006) has also been found in the nucleus and appears to be involved in DNA damage response via p53 signalling (Jung et al. 2006). Myosin 16b on the otherhand appears to be involved in DNA replication (Cameron et al. 2007) but it is not clear yet if this is in a direct way or through structural interactions with replication machinery (Castano et al. 2010). Myosin Va (Pranchevicius et al. 2008) and Myosin Vb (Lindsay and McCaffrey 2009a) are both found in the nucleus, specifically in nucleoli with myosin Va also being localised in nuclear speckles. These findings have increased interest in understanding nuclear myosin's involvement in processes such as transcription, chromatin remodelling and the movement of chromatin around nuclei (Hofmann et al. 2006a).

5.3.2.1 Nuclear Myosin 1β (NMIβ)

The gene MYO1C present on human chromosome 17p13, encodes three myosin isoforms including myosin 1C a, b and c. Out of the three isoforms, myosin 1C isoform b, also known as myosin 1β is the shortest isoform, has a unique N-terminal domain as compared to the other isoforms is 120 kDa, is evolutionarily conserved in vertebrates (Kahle et al. 2007) but is probably present in simpler invertebrate organisms (Hofmann et al. 2009) and is nuclear (Nowak et al. 1997). Nuclear myosin 1β (NM1β) is usually found as monomers within the cell nucleus (Roberts et al. 2004). At least four different sub-classes of myosin I have been identified, all of which, including NMIβ consists of a globular head and a single heavy chain (Gillespie et al. 2001) and possess the signature properties of a myosin 1 molecule including affinity for actin binding and K^+-EDTA ATPase activity (Nowak et al. 1997).

Distinguishing it from the cytoplasmic myosin 1 (CMI) molecule, NMIβ molecule has a unique 16-residue amino acid extension on its N-terminal end. These extra amino acid residues do not have a homology to nuclear localization sequence but cleaving this extension off the nuclear myosin 1 isoform (NMI) results in retention in the cytoplasm, thus signifying the importance of this extension for nuclear entry and retention of NM1 (Pestic-Dragovich et al. 2000).

NMIβ, when first observed by Nowak et al. was found to be evenly distributed in the nuclei throughout the nucleoplasm and was also observed to co-localise with nucleolar structures (Nowak et al. 1997) which was suggestive of possible involvement of this protein in transcribing ribosomal genes. Indeed, in a study performed by de Lanerolle's group, association of NMIβ with RNA polymerase I was identified in both in vivo and in vitro systems (Philimonenko et al. 2004). Further, this study demonstrated that inhibition of NMIβ using RNA interference technique led to production of pre-rRNA in HeLa cells, while over-expression of this protein increased the pre-rRNA production in a dose-dependent manner (Philimonenko et al. 2004). In addition to this, the in vitro studies demonstrated that NMI was essential for the maximum activity of RNA polymerase I (Philimonenko et al. 2004). Interactions between NMI and RNA polymerase II by co-precipitation assays have also been identified (Pestic-Dragovich et al. 2000). In vivo and in vitro studies demonstrated that blocking NMI inhibits RNA polymerase II transcription (Pestic-Dragovich et al. 2000; Hofmann et al. 2006b). Further, by using an abortive transcription assay and a detailed analysis of different steps in early transcription, NMI was shown to be involved in formation of the first phosphodiester bond in RNA polymerase II mediated transcription (Hofmann et al. 2006b). Thus, the involvement of NMI in transcription by RNA polymerases I, II and III has been illustrated (Hu et al. 2008; Philimonenko et al. 2004; Pestic-Dragovich et al. 2000; Vreugde et al. 2006; Lindsay and McCaffrey 2009b; Fomproix and Percipalle 2004; Grummt 2003; Cavellan et al. 2006). Mass spectrometry and immunoprecipitation studies have allowed identification of interactions between NMI and WSTF-SNF2h components of chromatin remodelling complex B-WICH (Cavellan et al. 2006; Percipalle and Farrants 2006; Percipalle et al. 2006), suggesting a probable role of NMI in recruiting these chromatin remodelling complexes to rDNA.

5.3.2.2 Distribution of NM1β

Previous studies employing immunoelectron microscopy (Nowak et al. 1997) and indirect immunofluroescence (Fomproix and Percipalle 2004) have demonstrated the presence of NMIβ within dense fibrillar compartment and granular compartment of the nucleolus, where it has been shown to interact with RNA polymerase I (Philimonenko et al. 2004; Fomproix and Percipalle 2004; Percipalle et al. 2006; Kysela et al. 2005) and thus influences transcription (Philimonenko et al. 2004) In addition to this, NMIβ has also been documented to be located within regions condensed and decondensed chromatin and it also colocalises with transcription sites (Kysela et al. 2005). In control proliferating HDFs, NMIβ is indeed localised in the nucleolus. By performing dual staining experiments with pKi-67 which is known to localise in the DFC domain of the nucleoli, the presence of NMIβ within this nucelolar domain has also been confirmed. Whether or not NMIβ colocalise or interact with nucleolar proteins such as nucleolin or Ki-67 present in the same nucleolar domain still remains to be explored. In addition to the nucleolar distribution, NMIβ is also distributed at the nuclear rim by the nuclear envelope in proliferating HDFs as well as foci throughout the nucleoplasm. The functional relevance of presence of NMIβ at these nuclear sites still remains unclear, but hypotheses suggesting interaction between nuclear envelope protein, namely emerin and lamin A, and NMIβ has been put forward (Mehta et al. 2008; Holaska and Wilson 2007). NMIβ has been implicated in nucleocytoplasmic transport, hence it is a possibility that NMIβ may interact with components of nuclear pore complexes (Obrdlik et al. 2010) or with other INM and ONM proteins such as SUN domain proteins, nesprins, some of which have an affinity towards cytoplasmic myosins (Zhang et al. 2005). NMIβ, being a part of WICH (WSTF-SNF2h complex), and being involved in movement of chromatin regions (Dundr et al. 2007; Chuang et al. 2006) and whole chromosomes (Hu et al. 2008; Mehta et al. 2010) across the nucleus, the nucleoplasmic pool of NMIβ may be facilitating chromatin remodelling and intranuclear chromosomal repositioning (Fig. 5.3). NMIβ colocalises with nuclear actin in plant cell nuclei and is present in the nucleolus, with putative transcription foci and in an intranuclear network (Cruz et al. 2009).

The distribution of NM1β in nuclei is the distribution seen in proliferating since there are changes to the distribution in non-proliferating cells (Mehta et al. 2010; Kysela et al. 2005). The relevance this has to NM1β driven processes in non-proliferating cells is not yet clear (Fig. 5.4).

In lamin A mutant cell lines derived from Hutchinson-Gilford Progeria Syndrome patients, NMIβ is predominantly distributed in nucleoplasmic aggregates as observed in non-proliferating control cells. In these cells the mutant form of lamin A remains abnormally farnesylated. Treatment of these cells with farnesyl transferase inhibitors (FTI) restores NMIβ to the nuclear rim, nucleoplasm and nucleolus in a proliferating subset of cells. Thus lamin A is vital for organisation and maintenance of NMIβ within proliferating cell nucleus and the presence of farnesylated forms of lamin A interferes with distribution of NMIβ. The interaction of lamin A and NM1β may well be through a mutual binding partner emerin. Whether or not transcription is

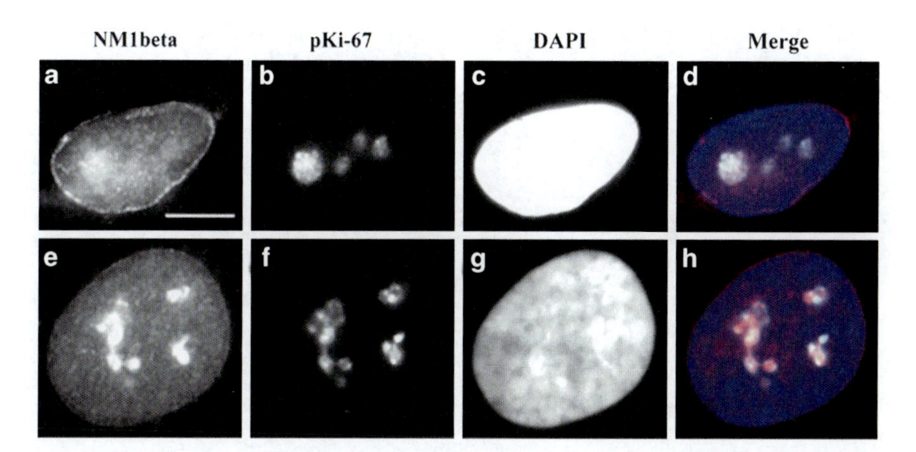

Fig. 5.3 The nuclear distribution of anti-nuclear myosin 1β in proliferating primary human dermal fibroblasts. Panels (**a**, **e**) display the anti-NM1β staining employing a commercial antibody (Mehta et al. 2010). The distribution is nucleolar, as determined by anti-pKi67 staining (**b**, **f**) which not only reveals nucleoli but also acts as a proliferation marker; at the periphery of the nucleus and throughout the nucleoplasm as homogenous/punctuate staining. Scale bar = 5 μm

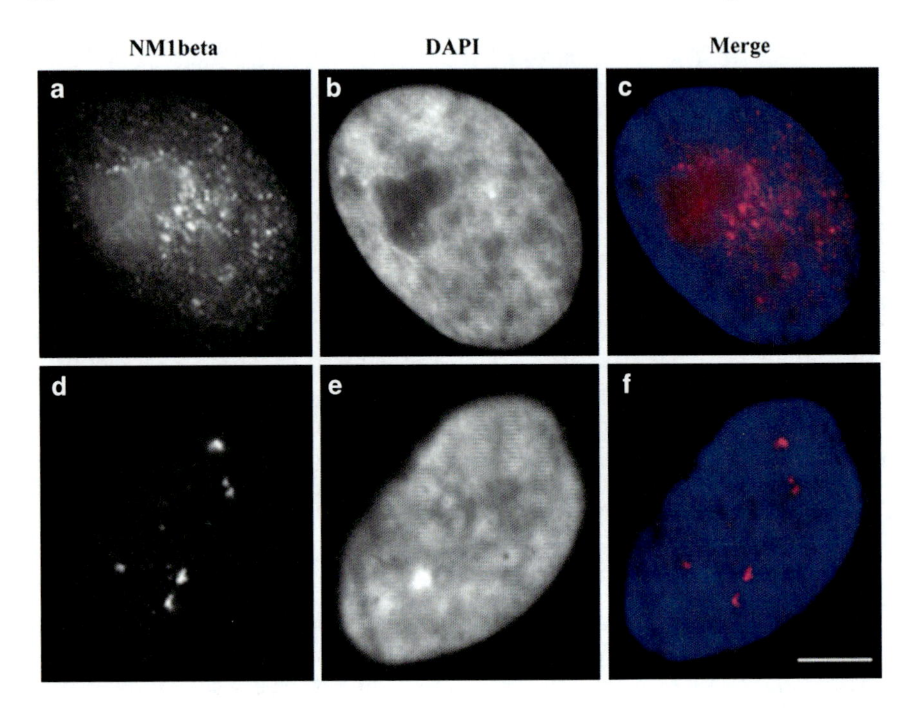

Fig. 5.4 The nuclear distribution of anti-nuclear myosin 1β in non-proliferating primary human dermal fibroblasts. In non-proliferating fibroblasts, as selected by the absence of pKi-67 staining in a dual colour indirect immunofluorescence, the nuclear distribution of anti-NM1β is altered with large aggregates towards the centre of the nucleus and a loss of the nucleolar and nuclear envelope staining. Scale bar = 5 μm

directly affected in HGPS cells due to disorganised NMIβ is unknown but active chromosome movement is restored in HGPS cells after FTI treatment (Mehta IS, Kill IR and Bridger JM manuscript in submission).

5.4 Nuclear Motor Protein Involvement in Long Range Movement of Chromatin

Although chromosomes within a cell nucleus are mostly constrained (Chubb et al. 2002; Sullivan and Shelby 1999; Weipoltshammer et al. 1999), fixed or anchored by the nuclear structural elements of the nucleus (Foster and Bridger 2005), there are instances when individual gene loci, chromosomal sub-regions or whole chromosomes move to new compartments of the nucleus. Most often this is in response to an external stimulus and is related to gene activation or silencing. The acceptance of this is fairly recent but was already apparent in the early part of this decade with respect to gene clusters coming away from whole chromosome territories as in the MHC complex on chromosome 6 (Volpi et al. 2000) whole individual chromosomes relocating within interphase nuclei upon serum removal or cells becoming senescent (Bridger et al. 2000), repositioning of genes in lymphocytes (Brown et al. 1999) and various regions of the genome being relocated during the cell cycle and after irradiation (Skalnikova et al. 2000). With the recognition that chromatin did indeed change locations, and for the MHC complex it was found to be rapid (~15 min) (Volpi et al. 2000), it took a while for the connection between the nuclear motor proteins presence in the nucleus and chromosome/gene movement to be validated.

5.4.1 Activated Gene Loci Movement in Real Time

Belmont and colleagues were able to analyse chromatin movements in real-time by creating regions of the genome that had incorporated a vector containing the lac operator. The cells also contain exogenous lac repressor protein engineered to not only contain a nuclear localisation signal (NLS) but also be fused to green fluorescent protein. Thus chromosomes can be tagged by the lac operator system and visualised in live cells through GFP-lac repressor (Belmont 2001) and references therein. When a response sequence as in the VP16 acidic activation domain was tethered to a single chromosomal site of lac operator elements transcriptional activation of this site the movement of the site from the nuclear periphery to the nuclear interior became permanent (Tumbar and Belmont 2001) (Fig. 5.5). The development of these impressive in vivo visualisation methodologies combined with two-photon excitation microscopy and understanding of biophysics (Levi et al. 2005), allowed Belmont and colleagues to ask questions of this directional movement and whether motor proteins were involved. Their paper in 2006 is really the first real milestone in interrogating chromosomal movements via active directed motor complex (Chuang et al. 2006). The authors adapted their GFP tagged chromosome system so that the addition of rapamycin to induce the VP16 acidic activation domain which resulted in the translocation of the chromosomal site to the nuclear interior, it did not matter whether the targeted synthetic peptide induced transcription or not, the site still relocated. Some of the GFP spots moved almost 3 μm over a 2 h time period but what was most interesting was the short bursts of very rapid and directed movement, where the movement was more like 1 μm/min. Indeed, the spots seemed to always follow curvilinear trajectories. Using an antibody against NM1β Belmont and colleagues found a focus of NM1β colocalising with the tagged chromosomal region. After transfection of a dominant negative mutant NM1β the repositioning of the rapamycin/VP16 locus was delayed. Locus repositioning was totally blocked with the addition of 2–3Butane-dionemonoxime (BDM) to the medium, which affects actin dynamics. This result was corroborated by the transfection of an actin mutant unable to polymerise, but with another actin mutant that permitted the stabilisation of F-actin the spot was able to relocalise. Thus, this pioneering work implicates NM1β in concert with polymerised actin i.e. F-actin in long range chromatin movements. Chuang and Belmont also make

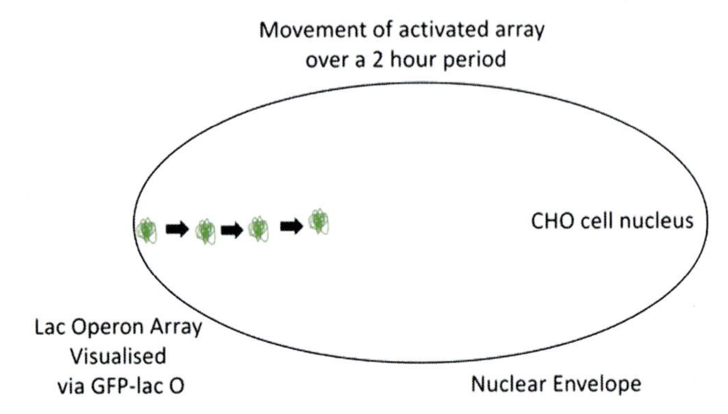

Fig. 5.5 Is a cartoon depiction of the results from the Chuang and Belmont study whereby they introduced arrays of the lac operator sequences into Chinese hamster ovary cells (CHO). The arrays are visualised in real-time studies by the association of GFP-lac repressor fusion protein with the lacO sequences. The arrays also contain a VP16 acidic activation domain so that gene expression can be activated. When this happens the fluorescently tagged arrays move into the nuclear interior, as depicted by the black arrows. This movement is not seen when the drugs that inhibit actin and myosin polymerisation are used or dominant negative mutant myosin or actin are expressed in the cells. The movement of the arrays can be very rapid at times implying active movement and they have a curvilinear trajectory implying the relocation is directed (Chuang and Belmont 2007)

the very pertinent point that since there could be active directed movement it does not preclude non-active diffusional movement (Chuang and Belmont 2007).

The nuclear positioning of ETO of AML 1 genes, that are often involved in translocation in acute myeloid leukaemia, is not proximal in human fibroblasts (Rubtsov et al. 2008). However when the cells are treated with etoposide, a cancer therapy, the two genes are much closer together. This repositioning is inhibited when the drug BDM is also used, again revealing motor activity associated with gene loci movement.

5.4.2 Exogenous Plasmid Intranuclear Movements Use the Host Cells' Nuclear Actin

In another set of experiments exogenous sequences have been used to study chromatin/DNA dynamics of liposomally-introduced plasmids (Ondrej et al. 2007; Ondrej et al. 2008a; Ondrej et al. 2008b). Using MCF7 cells the authors transfected into cells plasmids that were fluorescently tagged alongside a vector with a fusion gene for actin-Lumio. These clever transfections allowed the authors to visualise both the plasmid DNA and the actin network within cells at different fluorescent wavelengths in real-time. The fluorescent actin was found both in the cytoplasm but mainly in the nucleoplasm as large aggregates and more punctuate distributions, giving an impression of a comprehensive network that cannot be observed with antibody staining alone. When the fluorescently labelled plasmid DNA was also visualised in live cells they were found to associate with the actin at the nuclear periphery and in the aggregates throughout the nuclei. Double strand breaks (DSB) were induced by gamma-irradiation and visualised through H2AX staining after fixation. Interestingly the number of actin aggregates had increased and labelled plasmids had coalesced at the DSB. This implies that the plasmid DNA has travelled through the nucleoplasm and found the DSB. The authors then employed actin polymerisation inhibitor latrunculin B prior to irradiation and showed that the fraction of plasmids colocalised to DSBs was significantly decreased. These studies imply that movement of loci to repair factories requires an actin-based motor.

5.4.3 Rapid Repositioning of Whole Chromosome Territories

In the afore-mentioned review Chuang and Belmont (Chuang and Belmont 2007) discuss the repositioning of whole individual chromosome territories in cells that do not go through mitosis and a subsequent breakdown and re-association of chromosomal regions with nuclear structural elements. Thus, there are mechanisms in place that permit a non-random repositioning of whole chromosomes seen early on in the field's history in neuronal cells of the X chromosome (Barr and Bertram 1951; Borden and Manuelidis 1988).

Our interest in chromosome repositioning mechanisms was initiated with the observation that gene-poor chromosome 18, normally at the nuclear periphery in proliferating primary fibroblasts (Bridger et al. 2000; Croft et al. 1999; Meaburn et al. 2008) was located in the nuclear interior, associated with nucleoli, in cells that had been induced into quiescence by 7 days serum starvation or had become senescent in culture after serial passage for a number of months (Bridger et al. 2000). Later we also found other gene-poor chromosomes, such as chromosome 13 also relocated to the nuclear interior from the nuclear periphery when cells became senescent (Meaburn et al. 2007). Interestingly we never found any X chromosome repositioning from the nuclear periphery in any of our studies. It is not clear whether pre-senescent cells would relocate chromosome 18 with or without going through mitosis (Mehta et al. 2007) but it was possible to test this in the cells that had been placed in low serum. By analysing chromosome position shortly after serum removal from proliferating cultures we were able to see in "snap-shot" rapid repositioning of chromosomes 18 and 13 to the nuclear interior and chromosome 10 heading in the opposite direction from an intermediate location to a peripheral one. Again chromosome X territories did not move from the nuclear periphery (Mehta et al. 2008; Mehta et al. 2010). This is exciting for two different reasons – it means that individual chromosomes take different paths, or no path at all, when cells change their proliferative status and as such there must be a reason that needs to "fathomed out" and that these whole chromosome movements are directional and actually very rapid (Fig. 5.6). Repositioning took less than 15 min after the serum had been removed from the cultures. We believe that this chromosomal

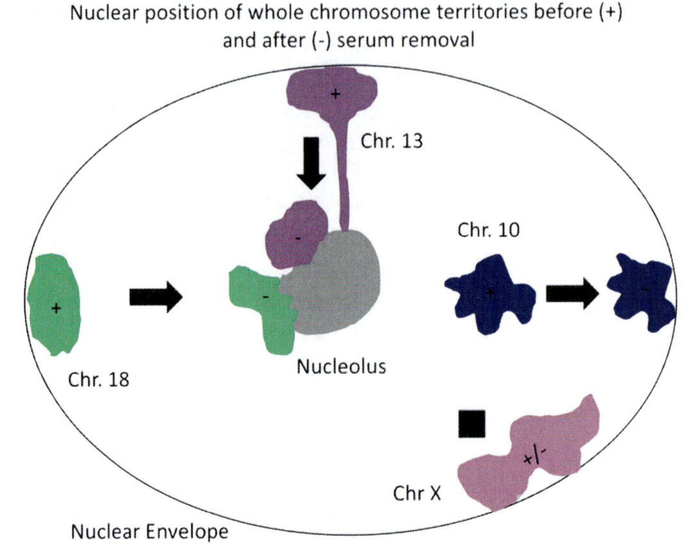

Fig. 5.6 Is a cartoon depicting the relocation of chromosome territories in primary human dermal fibroblasts after the removal of serum from the culture (Mehta et al. 2010). The +sign indicates a chromosome territory in a cell that is in 10% serum, a – sign is the territory when the serum has been removed from the culture for more than 15 min. Within 15 min chromosomes 18 and 13 have moved from the nuclear periphery to the nuclear interior and are found associated with the nucleolus. This is because chromosome 13 always has a link with the nucleolus through its ribosomal repeats, as is depicted by the elongated region of the territory with the + sign. Chromosome 10 moves in the opposite direction to chromosomes 13 and 18 from an intermediate position to the nuclear periphery and chromosome X does not change its nuclear location at all. This rapid repositioning is inhibited when drugs interfering with actin or myosin polymerisation are used or NM1β expression is interfered with using siRNA construct. This data imply that whole chromosomes are moved rapidly via a nuclear motor complex

repositioning assay is going to be an important tool in understanding chromosome dynamics and it needs to be adapted for real-time studies similar to Belmont's experiments. However, taking of 2D and 3D samples for fluorescence in situ hybridisation (FISH) (for methodologies see (Bridger JM 2010)) every few minutes does afford appreciation of dynamic movement. Establishing real-time studies will not be easy since the cells need to be serum responsive and we need to tag specific chromosomes. We are presently working on methodologies to do this in a sophisticated way, i.e., Fluorescence in vivo Hybridisation (Wiegant JB et al. 2010).

When comparing rates of chromosome movement we have similar rates to Belmont since the chromosomes can move 10–15 μm in 15 min, but these studies are looking at an averaged position from over 100 chromosome territories. However, the rapidity of chromosome reposition was impressive and again pointed towards active, directional movement, especially since there were chromosomes headed in different directions concomitantly. Following in Belmont's footsteps we used drugs that inhibited polymerisation of myosin (and actin) such as BDM, Jasplakinolide which halted the serum-responsive chromosome repositioning, which Latrunculin A also did. Interestingly, the inhibitor of actin polymerisation phalloidin oleate did not block the chromosome movement and serendipitously we had used it in such as way as it only affected cytoplasmic actin – thus adding credence to the hypothesis that whole chromosome movement is elicited by nuclear acto-myosin motor activity. We also used drugs, AG10 and ouabain to inhibit GTPase and ATPase activity, respectively – they also resulted in no chromosome relocation. Instead of transfecting dominant mutant actin or myosin genes we employed suppression of NM1 β expression using the MYO1C siRNA construct, transiently transfected – as used by (Philimonenko et al. 2004). This was so successful that >95% of fibroblasts no longer contained any discernable NM1β staining in the nucleus. We were not able to induce any shift in the chromosomes after this knock-down. We feel that these data are strong evidence for long range chromosomal movement being elicited by a motor complex similar to those seen in the cytoplasm for moving cargo. However, given that these proteins are also involved in chromatin modelling and modification – we need to look into whether this is the driving force behind individual chromosomes behaving differently to each other with respect to nuclear reposition.

5.4.4 *Active Chromosomal Movement Towards a Specific Nuclear Entity*

Both the studies discussed above describe active directional chromosomal or sub-chromosomal within nuclei but the destination and reason for the movement is not yet clear. However, other studies have demonstrated actin/myosin motor reliant movement of chromosomes and/or genes to specific nuclear structures/bodies. The first study is from the Matera laboratory and they have employed 4D imaging of U2 snRNA genes and their subsequent association with Cajal bodies correlated with their expression, although the actual reason for the colocalisation is not yet elucidated (Dundr et al. 2007). These authors worked in 4D and constructed cells that contained U2 snRNA arrays with the lac operator repeats as in Belmont's study. The array could then be visualised via mCherry lac repressor. The cells were also transfected with a vector containing a Cajal bodies marker protein (coilin) fused to GFP. Both entities could then be followed in real-time in 3D in two different fluorescent colours. For photosensitivity reasons used a spinning disk confocal microscope. After induction of the U2 snRNA exogenous genes there were temporary collisions between the Cajal bodies and the gene arrays with tight and lasting colocalisation apparent after 6–7 h with the time just before the final more lasting collision revealing more active movement of the U2 genes towards the Cajal bodies, as calculated by a bespoke targeted component tracking analysis. The U2 arrays that were associated with Cajal bodies were expressing as revealed by RNA-FISH. DNA-FISH with probes for the U2 snRNA gene arrays and the chromosome they were integrated upon revealed that when activated the array would often be found on a chromatin loop emanating away from the chromosome territory. Since this movement of genes from the chromosome territory towards a nuclear body appeared directed and active, the authors used a similar approach to Belmont by expressing an actin mutant that prevent actin polymerisation. The results were that neither the chromatin loops were formed nor gene array:Cajal body association seen.

In another study, Hu et al. (2008) revealed a number of different chromosome territories relocating to interchromatin granules in MCF7 or human mammary epithelial cells that responded to 17b-estradiol. The authors used chromosome conformation capture (3C) and a specialised chromatin immunopreciptation assay to assess the level of interaction between ERa- bound activated genes, such as *GREB1* and the 17b-estradiol regulated *TFF1* gene. Using 3D-FISH these authors found that after 17β-estradiol treatment the two chromosomes housing the two genes were now neighbours. Although these authors have performed no timed experiments for the colocalisation of the 17β-estradiol activated genes and their chromosomes when the cells were treated with Latrunculin or Jasplakinolide the genes and chromosomes no longer became intimately situated in interphase nuclei. In addition to the drug treatment the authors also treated their cells with siRNA constructs and injected antibodies for NM1β and G-actin, also resulting in no responsive repositioning. Further they also employed mutant constructs of NM1β that also had the same effect. The drugs also prevented the gene association with interchromatin granules in 17β-estradiol treated cells. Another study, recently published, has shown that these genes in similar cells do not colocalise upon activation with estradiol (Kocanova et al. 2010).

5.4.5 Long Range Chromatin Movement to the Nuclear Periphery in Yeast

In yeast nuclear biology, a wide range of genes have been found to relocate to the nuclear periphery after activation (Brickner and Walter 2004), associated with nuclear pores (Casolari et al. 2004; Taddei et al. 2006), which is oppositional to what appears to happen on the whole in vertebrate cells i.e. activated genes tend to move towards the nuclear interior, if they relocate. A number of the genes in yeast that move to the nuclear periphery also attach to nuclear pores proteins via a myosin-like protein Mlp1 and this aids in the transcriptional memory i.e. allows their re-expression to be elicited more rapidly than if they were not anchored at the nuclear pore (Tan-Wong et al. 2009; Laine et al. 2009). Although, this is a very exciting model system of positional control of a genome for the purposes of this chapter it is the work by (Yoshida et al. 2010) that has demonstrated a possible motor function in these genes reaching the nuclear pores. The budding yeast actin-related protein – Arp6 – apart from its role as a transcriptionally relevant remodelling of gene promoters, appears to have a further role in long range movement of ribosomal genes to the nuclear periphery (Yoshida et al. 2010). In order to substantiate this role for Arp6 the authors used an interesting assay that permits the determination of a specific candidate protein fused to LexA i.e Arp6, to move a randomly located chromosomal locus to the nuclear periphery. The random site contains LexA binding sites as well as the lac operator array allowing real-time visualisation via GFP lac I. When Arp6 was fused to LexA the chromosomal loci were found more abundantly located at the edge of the nucleus. When performed in a mutant cell-line where functional pores are only found on one side of the nucleus – the Arp6-LexA bound loci were located with the nuclear pores. By using ChIP analysis in Apr6 mutant cells with an antibody to nuclear pore protein Nup133 the authors also found other genes normally targeted to nuclear pores were not. Association of genes with the nuclear pores was also lost in Mlp1 (and Mlp2) mutant cells, although cell cycle stage specific these data imply that in yeast there exists a form of acto-myosin activity that is responsible for moving genes across the nucleus (Dion et al. 2010).

5.5 Future Perspectives

We have described all the studies so far that have implicated nuclear motor activity in the long range movement of chromatin through the nucleoplasm. The organisms so far studied are budding yeast, CHO (hamster) cells and the human cells which is a very wide spectrum and the gap between these organisms needs to be closed by other models, especially in those where manipulation of the genome and knock-in and knock-outs are relatively straight-forward. It would also be necessary that their genome structure and behaviour had already been mapped out to a certain extent. Such organisms could include *Drosophila*, given the amount of pioneering work that has been performed on the dynamics of the *Drosophila* chromosomes. Other important model organisms for study could be *C. elegans* and Zebra fish, since they are good models to use live because of the possibility of being able to see cells and sub-cellular real-time dynamics because of their transparency, especially during development. However, more would need to be known about the spatial and temporal behaviour of their genes, chromosomes and genome and nuclear motor candidate genes indentified.

Some of the studies described have used real-time fluorescence imaging to follow the dynamics of their locus/chromosome-tag of interest and have created complex of being able to do multi-colour experiments. Indeed Kozubek and colleagues attempted to visualise both the chromatin structure and a component of a nuclear motor in the same cells. This would be the goal to see both elements i.e. "the cart and the horse" but in combination with these kinds of live cell imaging there needs to be built in quantitation of distance moved, rate of movement, direction and pathway and perhaps even the possibility to see the rest of the nuclear landscape. These studies also need to be performed in large numbers in order to quantitate securely a general mechanism. One could envisage that these types of studies should become automated and may even provide diagnostic value – given that in some disease gene position within the interphase nucleus is altered. Furthermore, masterminds of studies using fluorescent tags and exogenous loci or altered chromatin states and amplification as a result of tagging a site for visualisation need to consider the effect changing their chromosomal site of choice will have on the dynamics and interactions with nuclear structure and motor proteins.

The involvement of enigmatic structures like nuclear motors in chromatin/chromosome dynamics needs to be modelled *in silico* very carefully intergrating computer scientists, systems biologists and cell biologists. It would seem pertinent to include nuclear and genome biologists encompassing a wide-range of knowledge of the cell nucleus at different levels of organisation i.e. detailed behaviour of specific genes, those with knowledge of specific nuclear bodies and structures and those with an overview of spatial and temporal nuclear dynamics. This will be difficult and but not impossible. One of the major stumbling blocks will be the inability of the field to as yet decide definitively if there exists a nuclear matrix/structure and if so how it is organised and regulated in different states of proliferation, cell types, organisms and point in development or ageing. The lack of major progress in this area is due to how difficult it is to visualise structures, potentially networks deep within the nucleoplasm and the field needs to develop even better microscopy and imaging manipulation and analysis and biochemical but physiologically relevant protocols to reveal these structures. It will be critical that these structures can be visualised in action in real-time and we need to overcome problems with fluorescence toxicity and the alteration of the delicate balance of chromatin modification/modelling at targeted genomic sites.

Basic modelling has been done with respect to how an acto-myosin might work in the nucleus to actively move chromatin/chromosomes (Hofmann et al. 2006a). This model suggests that NMIβ could bind via its tail to the nuclear entity that needs to be transported around and then actin binds to the globular head of the NMIβ, and this nuclear motor like cytoplasmic actin-myosin motor would then translocate the nuclear entity along the tracks of highly dynamic nuclear actin (Hofmann and de Lanerolle 2006). However, whether the association of actin and myosin is required for their role in transcription or if they work independently is still unclear. It would be of interest to understand what signalling mechanism is involved in the active nuclear motor dependent movement of chromosomes around the interphase nucleus. Whether or not these signals translated from the cytoplasm involving SUN proteins and KASH domain proteins (Chikashige et al. 2007; Haque et al. 2006; Starr 2009) and involve structural proteins such as emerin and lamin A stills need to be resolved. The field will also need to put in place large scale protocols for identifying in vivo protein:protein interactions and chromatin:protein interactions with respect to nuclear motor proteins.

Understanding whether actin and nuclear myosins have cargo transportation roles in the nucleus is a critical step in our understanding of how the genome functions in time and space and fundamental task for this century in fully understanding our genomes in health and disease.

Acknowledgements We would like to thank the Brunel University Progeria Research Fund for partially funding ISM in her Ph.D. studies.

References

Pollard TD, Eisenberg E, Korn ED, Kielley WW (1973) Inhibition of Mg ++ ATPase activity of actin-activated Acanthamoeba myosin by muscle troponin-tropomyosin: implications for the mechanism of control of amoeba motility and muscle contraction. Biochem Biophys Res Commun 51:693–698

Pollard TD, Korn ED (1973) Acanthamoeba myosin. II. Interaction with actin and with a new cofactor protein required for actin activation of Mg 2+ adenosine triphosphatase activity. J Biol Chem 248:4691–4697

Berg JS, Powell BC, Cheney RE (2001) A millennial myosin census. Mol Biol Cell 12:780–794

Bahler M (2000) Are class III and class IX myosins motorized signalling molecules? Biochim Biophys Acta 1496:52–59

Yin H, Pruyne D, Huffaker TC, Bretscher A (2000) Myosin V orientates the mitotic spindle in yeast. Nature 406:1013–1015

Evangelista M, Klebl BM, Tong AH, Webb BA, Leeuw T et al (2000) A role for myosin-I in actin assembly through interactions with Vrp1p, Bee1p, and the Arp2/3 complex. J Cell Biol 148:353–362

Lechler T, Shevchenko A, Li R (2000) Direct involvement of yeast type I myosins in Cdc42-dependent actin polymerization. J Cell Biol 148:363–373

Lee WL, Bezanilla M, Pollard TD (2000) Fission yeast myosin-I, Myo1p, stimulates actin assembly by Arp2/3 complex and shares functions with WASp. J Cell Biol 151:789–800

Sellers JR (2000) Myosins: a diverse superfamily. Biochim Biophys Acta 1496:3–22

Mermall V, Post PL, Mooseker MS (1998) Unconventional myosins in cell movement, membrane traffic, and signal transduction. Science 279:527–533

McDonald D, Carrero G, Andrin C, de Vries G, Hendzel MJ (2006) Nucleoplasmic beta-actin exists in a dynamic equilibrium between low-mobility polymeric species and rapidly diffusing populations. J Cell Biol 172:541–552

de Lanerolle P, Johnson T, Hofmann WA (2005) Actin and myosin I in the nucleus: what next? Nat Struct Mol Biol 12:742–746

Hofmann WA, Johnson T, Klapczynski M, Fan JL, de Lanerolle P (2006a) From transcription to transport: emerging roles for nuclear myosin I. Biochem Cell Biol 84:418–426

Castano E, Philimonenko VV, Kahle M, Fukalova J, Kalendova A et al (2010) Actin complexes in the cell nucleus: new stones in an old field. Histochem Cell Biol 133:607–626

Dion V, Shimada K, Gasser SM (2010) Actin-related proteins in the nucleus: life beyond chromatin remodelers. Curr Opin Cell Biol

Branco MR, Pombo A (2006) Intermingling of chromosome territories in interphase suggests role in translocations and transcription-dependent associations. PLoS Biol 4:e138

Osborne CS, Chakalova L, Mitchell JA, Horton A, Wood AL et al (2007) Myc dynamically and preferentially relocates to a transcription factory occupied by Igh. PLoS Biol 5:e192

Dundr M, Ospina JK, Sung MH, John S, Upender M et al (2007) Actin-dependent intranuclear repositioning of an active gene locus in vivo. J Cell Biol 179:1095–1103

Chuang CH, Carpenter AE, Fuchsova B, Johnson T, de Lanerolle P et al (2006) Long-range directional movement of an interphase chromosome site. Curr Biol 16:825–831

Mehta IS (2005) Genome organisation in senescent cells. Master's dissertation

Mehta IS, Elcock LS, Amira M, Kill IR, Bridger JM (2008) Nuclear motors and nuclear structures containing A-type lamins and emerin: is there a functional link? Biochem Soc Trans 36:1384–1388

Hu Q, Kwon YS, Nunez E, Cardamone MD, Hutt KR et al (2008) Enhancing nuclear receptor-induced transcription requires nuclear motor and LSD1-dependent gene networking in interchromatin granules. Proc Natl Acad Sci U S A 105:19199–19204

Mehta IS, Amira M, Harvey AJ, Bridger JM (2010) Rapid chromosome territory relocation by nuclear motor activity in response to serum removal in primary human fibroblasts. Genome Biol 11:R5

Lane NJ (1969) Intranuclear fibrillar bodies in actinomycin D-treated oocytes. J Cell Biol 40:286–291
Clark TG, Rosenbaum JL (1979) An actin filament matrix in hand-isolated nuclei of X. laevis oocytes. Cell 18:1101–1108
Nakayasu H, Ueda K (1983) Association of actin with the nuclear matrix from bovine lymphocytes. Exp Cell Res 143:55–62
Jockusch BM, Becker M, Hindennach I, Jockusch E (1974) Slime mould actin: homology to vertebrate actin and presence in the nucleus. Exp Cell Res 89:241–246
Clark TG, Merriam RW (1977) Diffusible and bound actin nuclei of Xenopus laevis oocytes. Cell 12:883–891
Egly JM, Miyamoto NG, Moncollin V, Chambon P (1984) Is actin a transcription initiation factor for RNA polymerase B? Embo J 3:2363–2371
Rungger D, Rungger-Brandle E, Chaponnier C, Gabbiani G (1979) Intranuclear injection of anti-actin antibodies into Xenopus oocytes blocks chromosome condensation. Nature 282:320–321
Scheer U, Hinssen H, Franke WW, Jockusch BM (1984) Microinjection of actin-binding proteins and actin antibodies demonstrates involvement of nuclear actin in transcription of lampbrush chromosomes. Cell 39:111–122
Ankenbauer T, Kleinschmidt JA, Walsh MJ, Weiner OH, Franke WW (1989) Identification of a widespread nuclear actin binding protein. Nature 342:822–825
Zhao K, Wang W, Rando OJ, Xue Y, Swiderek K et al (1998) Rapid and phosphoinositol-dependent binding of the SWI/SNF-like BAF complex to chromatin after T lymphocyte receptor signaling. Cell 95:625–636
Gonsior SM, Platz S, Buchmeier S, Scheer U, Jockusch BM et al (1999) Conformational difference between nuclear and cytoplasmic actin as detected by a monoclonal antibody. J Cell Sci 112(Pt 6):797–809
Schoenenberger CA, Buchmeier S, Boerries M, Sutterlin R, Aebi U et al (2005) Conformation-specific antibodies reveal distinct actin structures in the nucleus and the cytoplasm. J Struct Biol 152:157–168
Wang IF, Chang HY, Shen CK (2006) Actin-based modeling of a transcriptionally competent nuclear substructure induced by transcription inhibition. Exp Cell Res 312:3796–3807
Cruz JR, Diaz M, de la Espina S (2009) Subnuclear compartmentalization and function of actin and nuclear myosin I in plants. Chromosoma 118:193–207
Hofmann W, Reichart B, Ewald A, Muller E, Schmitt I et al (2001) Cofactor requirements for nuclear export of Rev response element (RRE)- and constitutive transport element (CTE)-containing retroviral RNAs. An unexpected role for actin. J Cell Biol 152:895–910
Kiseleva E, Drummond SP, Goldberg MW, Rutherford SA, Allen TD et al (2004) Actin- and protein-4.1-containing filaments link nuclear pore complexes to subnuclear organelles in Xenopus oocyte nuclei. J Cell Sci 117:2481–2490
Bohnsack MT, Stuven T, Kuhn C, Cordes VC, Gorlich D (2006) A selective block of nuclear actin export stabilizes the giant nuclei of Xenopus oocytes. Nat Cell Biol 8:257–263
Jockusch BM, Schoenenberger CA, Stetefeld J, Aebi U (2006) Tracking down the different forms of nuclear actin. Trends Cell Biol 16:391–396
Egelman EH (2003) A tale of two polymers: new insights into helical filaments. Nat Rev Mol Cell Biol 4:621–630
Gedge LJ, Morrison EE, Blair GE, Walker JH (2005) Nuclear actin is partially associated with Cajal bodies in human cells in culture and relocates to the nuclear periphery after infection of cells by adenovirus 5. Exp Cell Res 303:229–239
Muratani M, Gerlich D, Janicki SM, Gebhard M, Eils R et al (2002) Metabolic-energy-dependent movement of PML bodies within the mammalian cell nucleus. Nat Cell Biol 4:106–110
Olave IA, Reck-Peterson SL, Crabtree GR (2002) Nuclear actin and actin-related proteins in chromatin remodeling. Annu Rev Biochem 71:755–781
Pederson T (2000) Half a century of "the nuclear matrix". Mol Biol Cell 11:799–805
Zhang S, Buder K, Burkhardt C, Schlott B, Gorlach M et al (2002) Nuclear DNA helicase II/RNA helicase A binds to filamentous actin. J Biol Chem 277:843–853

Hofmann WA, Stojiljkovic L, Fuchsova B, Vargas GM, Mavrommatis E et al (2004) Actin is part of pre-initiation complexes and is necessary for transcription by RNA polymerase II. Nat Cell Biol 6:1094–1101

Philimonenko VV, Zhao J, Iben S, Dingova H, Kysela K et al (2004) Nuclear actin and myosin I are required for RNA polymerase I transcription. Nat Cell Biol 6:1165–1172

Hu P, Wu S, Hernandez N (2004) A role for beta-actin in RNA polymerase III transcription. Genes Dev 18:3010–3015

Kukalev A, Nord Y, Palmberg C, Bergman T, Percipalle P (2005) Actin and hnRNP U cooperate for productive transcription by RNA polymerase II. Nat Struct Mol Biol 12:238–244

Percipalle P, Jonsson A, Nashchekin D, Karlsson C, Bergman T et al (2002) Nuclear actin is associated with a specific subset of hnRNP A/B-type proteins. Nucleic Acids Res 30:1725–1734

Percipalle P, Zhao J, Pope B, Weeds A, Lindberg U et al (2001) Actin bound to the heterogeneous nuclear ribonucleoprotein hrp36 is associated with Balbiani ring mRNA from the gene to polysomes. J Cell Biol 153:229–236

Kimura T, Hashimoto I, Yamamoto A, Nishikawa M, Fujisawa JI (2000) Rev-dependent association of the intron-containing HIV-1 gag mRNA with the nuclear actin bundles and the inhibition of its nucleocytoplasmic transport by latrunculin-B. Genes Cells 5:289–307

Lee K, Song K (2007) Actin dysfunction activates ERK1/2 and delays entry into mitosis in mammalian cells. Cell Cycle 6:1487–1495

Andrin C, Hendzel MJ (2004) F-actin-dependent insolubility of chromatin-modifying components. J Biol Chem 279:25017–25023

Gieni RS, Hendzel MJ (2008) Mechanotransduction from the ECM to the genome: are the pieces now in place? J Cell Biochem 104:1964–1987

Gieni RS, Hendzel MJ (2009) Actin dynamics and functions in the interphase nucleus: moving toward an understanding of nuclear polymeric actin. Biochem Cell Biol 87:283–306

Holaska JM, Kowalski AK, Wilson KL (2004) Emerin caps the pointed end of actin filaments: evidence for an actin cortical network at the nuclear inner membrane. PLoS Biol 2:E231

Holaska JM, Wilson KL (2007) An emerin "proteome": purification of distinct emerin-containing complexes from HeLa cells suggests molecular basis for diverse roles including gene regulation, mRNA splicing, signaling, mechanosensing, and nuclear architecture. Biochemistry 46:8897–8908

Lee K, Kang MJ, Kwon SJ, Kwon YK, Kim KW et al (2007) Expansion of chromosome territories with chromatin decompaction in BAF53-depleted interphase cells. Mol Biol Cell 18:4013–4023

Sasseville AM, Langelier Y (1998) In vitro interaction of the carboxy-terminal domain of lamin A with actin. FEBS Lett 425:485–489

Dingova H, Fukalova J, Maninova M, Philimonenko VV, Hozak P (2009) Ultrastructural localization of actin and actin-binding proteins in the nucleus. Histochem Cell Biol 131:425–434

Hagen SJ, Kiehart DP, Kaiser DA, Pollard TD (1986) Characterization of monoclonal antibodies to Acanthamoeba myosin-I that cross-react with both myosin-II and low molecular mass nuclear proteins. J Cell Biol 103:2121–2128

Berrios M, Fisher PA, Matz EC (1991) Localization of a myosin heavy chain-like polypeptide to Drosophila nuclear pore complexes. Proc Natl Acad Sci U S A 88:219–223

Pederson T, Aebi U (2002) Actin in the nucleus: what form and what for? J Struct Biol 140:3–9

Nowak G, Pestic-Dragovich L, Hozak P, Philimonenko A, Simerly C et al (1997) Evidence for the presence of myosin I in the nucleus. J Biol Chem 272:17176–17181

Pestic-Dragovich L, Stojiljkovic L, Philimonenko AA, Nowak G, Ke Y et al (2000) A myosin I isoform in the nucleus. Science 290:337–341

Kahle M, Pridalova J, Spacek M, Dzijak R, Hozak P (2007) Nuclear myosin is ubiquitously expressed and evolutionary conserved in vertebrates. Histochem Cell Biol 127:139–148

Vreugde S, Ferrai C, Miluzio A, Hauben E, Marchisio PC et al (2006) Nuclear myosin VI enhances RNA polymerase II-dependent transcription. Mol Cell 23:749–755

Jung EJ, Liu G, Zhou W, Chen X (2006) Myosin VI is a mediator of the p53-dependent cell survival pathway. Mol Cell Biol 26:2175–2186

Cameron RS, Liu C, Mixon AS, Pihkala JP, Rahn RJ et al (2007) Myosin16b: The COOH-tail region directs localization to the nucleus and overexpression delays S-phase progression. Cell Motil Cytoskeleton 64:19–48

Pranchevicius MC, Baqui MM, Ishikawa-Ankerhold HC, Lourenco EV, Leao RM et al (2008) Myosin Va phosphorylated on Ser1650 is found in nuclear speckles and redistributes to nucleoli upon inhibition of transcription. Cell Motil Cytoskeleton 65:441–456

Lindsay AJ, McCaffrey MW (2009a) Myosin Vb localises to nucleoli and associates with the RNA polymerase I transcription complex. Cell Motil Cytoskeleton 66:1057–1072

Hofmann WA, Richards TA, de Lanerolle P (2009) Ancient animal ancestry for nuclear myosin. J Cell Sci 122:636–643

Roberts R, Lister I, Schmitz S, Walker M, Veigel C et al (2004) Myosin VI: cellular functions and motor properties. Philos Trans R Soc Lond B Biol Sci 359:1931–1944

Gillespie PG, Albanesi JP, Bahler M, Bement WM, Berg JS et al (2001) Myosin-I nomenclature. J Cell Biol 155:703–704

Hofmann WA, Vargas GM, Ramchandran R, Stojiljkovic L, Goodrich JA et al (2006b) Nuclear myosin I is necessary for the formation of the first phosphodiester bond during transcription initiation by RNA polymerase II. J Cell Biochem 99:1001–1009

Lindsay AJ, McCaffrey MW (2009) Myosin Vb localises to nucleoli and associates with the RNA polymerase I transcription complex. Cell Motil Cytoskeleton

Fomproix N, Percipalle P (2004) An actin-myosin complex on actively transcribing genes. Exp Cell Res 294:140–148

Grummt I (2003) Life on a planet of its own: regulation of RNA polymerase I transcription in the nucleolus. Genes Dev 17:1691–1702

Cavellan E, Asp P, Percipalle P, Farrants AK (2006) The WSTF-SNF2h chromatin remodeling complex interacts with several nuclear proteins in transcription. J Biol Chem 281:16264–16271

Percipalle P, Farrants AK (2006) Chromatin remodelling and transcription: be-WICHed by nuclear myosin 1. Curr Opin Cell Biol 18:267–274

Percipalle P, Fomproix N, Cavellan E, Voit R, Reimer G et al (2006) The chromatin remodelling complex WSTF-SNF2h interacts with nuclear myosin 1 and has a role in RNA polymerase I transcription. EMBO Rep 7:525–530

Kysela K, Philimonenko AA, Philimonenko VV, Janacek J, Kahle M et al (2005) Nuclear distribution of actin and myosin I depends on transcriptional activity of the cell. Histochem Cell Biol 124:347–358

Obrdlik A, Louvet E, Naschekin D, Kiseleva E, Fahrenkrog B et al (2010) Nuclear myosin 1 is in a complex with mature rRNA transcripts and associates with the nuclear pore basket. FASEB J 24:146–157

Zhang Q, Ragnauth CD, Skepper JN, Worth NF, Warren DT et al (2005) Nesprin-2 is a multi-isomeric protein that binds lamin and emerin at the nuclear envelope and forms a subcellular network in skeletal muscle. J Cell Sci 118:673–687

Chubb JR, Boyle S, Perry P, Bickmore WA (2002) Chromatin motion is constrained by association with nuclear compartments in human cells. Curr Biol 12:439–445

Sullivan KF, Shelby RD (1999) Using time-lapse confocal microscopy for analysis of centromere dynamics in human cells. Methods Cell Biol 58:183–202

Weipoltshammer K, Schofer C, Almeder M, Philimonenko VV, Frei K et al (1999) Intranuclear anchoring of repetitive DNA sequences: centromeres, telomeres, and ribosomal DNA. J Cell Biol 147:1409–1418

Foster HA, Bridger JM (2005) The genome and the nucleus: a marriage made by evolution. Genome organisation and nuclear architecture. Chromosoma 114:212–229

Volpi EV, Chevret E, Jones T, Vatcheva R, Williamson J et al (2000) Large-scale chromatin organization of the major histocompatibility complex and other regions of human chromosome 6 and its response to interferon in interphase nuclei. J Cell Sci 113(Pt 9):1565–1576

Bridger JM, Boyle S, Kill IR, Bickmore WA (2000) Re-modelling of nuclear architecture in quiescent and senescent human fibroblasts. Curr Biol 10:149–152

Brown KE, Baxter J, Graf D, Merkenschlager M, Fisher AG (1999) Dynamic repositioning of genes in the nucleus of lymphocytes preparing for cell division. Mol Cell 3:207–217

Skalnikova M, Kozubek S, Lukasova E, Bartova E, Jirsova P et al (2000) Spatial arrangement of genes, centromeres and chromosomes in human blood cell nuclei and its changes during the cell cycle, differentiation and after irradiation. Chromosome Res 8:487–499

Belmont AS (2001) Visualizing chromosome dynamics with GFP. Trends Cell Biol 11:250–257

Tumbar T, Belmont AS (2001) Interphase movements of a DNA chromosome region modulated by VP16 transcriptional activator. Nat Cell Biol 3:134–139

Levi V, Ruan Q, Plutz M, Belmont AS, Gratton E (2005) Chromatin dynamics in interphase cells revealed by tracking in a two-photon excitation microscope. Biophys J 89:4275–4285

Chuang CH, Belmont AS (2007) Moving chromatin within the interphase nucleus-controlled transitions? Semin Cell Dev Biol 18:698–706

Rubtsov MA, Terekhov SM, Razin SV, Iarovaia OV (2008) Repositioning of ETO gene in cells treated with VP-16, an inhibitor of DNA-topoisomerase II. J Cell Biochem 104:692–699

Ondrej V, Lukasova E, Falk M, Kozubek S (2007) The role of actin and microtubule networks in plasmid DNA intracellular trafficking. Acta Biochim Pol 54:657–663

Ondrej V, Lukasova E, Krejci J, Matula P, Kozubek S (2008a) Lamin A/C and polymeric actin in genome organization. Mol Cells 26:356–361

Ondrej V, Lukasova E, Krejci J, Kozubek S (2008b) Intranuclear trafficking of plasmid DNA is mediated by nuclear polymeric proteins lamins and actin. Acta Biochim Pol 55:307–315

Barr ML, Bertram EG (1951) The behaviour of nuclear structures during depletion and restoration of Nissl material in motor neurons. J Anat 85:171–181

Borden J, Manuelidis L (1988) Movement of the X chromosome in epilepsy. Science 242:1687–1691

Croft JA, Bridger JM, Boyle S, Perry P, Teague P et al (1999) Differences in the localization and morphology of chromosomes in the human nucleus. J Cell Biol 145:1119–1131

Meaburn KJ, Newbold RF, Bridger JM (2008) Positioning of human chromosomes in murine cell hybrids according to synteny. Chromosoma 117:579–591

Meaburn KJ, Cabuy E, Bonne G, Levy N, Morris GE et al (2007) Primary laminopathy fibroblasts display altered genome organization and apoptosis. Aging Cell 6:139–153

Mehta IS, Figgitt M, Clements CS, Kill IR, Bridger JM (2007) Alterations to nuclear architecture and genome behavior in senescent cells. Ann N Y Acad Sci 1100:250–263

Bridger JM Volpi (2010) Fluorescence in situ Hybridization (FISH), Molecular methods in molecular biology. Humana Press, USA

Wiegant JB, Raap AK; Tanke HJ, Dirks RW (2010) Visualizing nucleic acids in living cells by fluorescence in vivo hybridization. In: Bridger JM, Volpi EV (eds) Fluorescence in situ Hybridization (FISH), Methods in molecular biology. Humana Press, USA

Kocanova S, Kerr EA, Rafique S, Boyle S, Katz E et al (2010) Activation of estrogen-responsive genes does not require their nuclear co-localization. PLoS Genet 6:e1000922

Brickner JH, Walter P (2004) Gene recruitment of the activated INO1 locus to the nuclear membrane. PLoS Biol 2:e342

Casolari JM, Brown CR, Komili S, West J, Hieronymus H et al (2004) Genome-wide localization of the nuclear transport machinery couples transcriptional status and nuclear organization. Cell 117:427–439

Taddei A, Van Houwe G, Hediger F, Kalck V, Cubizolles F et al (2006) Nuclear pore association confers optimal expression levels for an inducible yeast gene. Nature 441:774–778

Tan-Wong SM, Wijayatilake HD, Proudfoot NJ (2009) Gene loops function to maintain transcriptional memory through interaction with the nuclear pore complex. Genes Dev 23: 2610–2624

Laine JP, Singh BN, Krishnamurthy S, Hampsey M (2009) A physiological role for gene loops in yeast. Genes Dev 23:2604–2609

Yoshida T, Shimada K, Oma Y, Kalck V, Akimura K et al (2010) Actin-related protein Arp6 influences H2A.Z-dependent and -independent gene expression and links ribosomal protein genes to nuclear pores. PLoS Genet 6:e1000910

Hofmann WA, de Lanerolle P (2006) Nuclear actin: to polymerize or not to polymerize. J Cell Biol 172:495–496

Chikashige Y, Haraguchi T, Hiraoka Y (2007) Another way to move chromosomes. Chromosoma 116:497–505

Haque F, Lloyd DJ, Smallwood DT, Dent CL, Shanahan CM et al (2006) SUN1 interacts with nuclear lamin A and cytoplasmic nesprins to provide a physical connection between the nuclear lamina and the cytoskeleton. Mol Cell Biol 26:3738–3751

Starr DA (2009) A nuclear-envelope bridge positions nuclei and moves chromosomes. J Cell Sci 122:577–586

Chapter 6
Methodology for Quantitative Analysis of 3-D Nuclear Architecture

Richard A. Russell, Niall M. Adams, David Stephens, Elizabeth Batty, Kirsten Jensen, and Paul S. Freemont

Abstract In the past 20 years cell biologists have studied the cell nucleus extensively, aided by advances in cell imaging technology and microscopy. Consequently, the volume of image data of the cell nucleus – and the compartments it contains – is growing rapidly. The spatial organisation of these nuclear compartments is thought to be fundamentally associated with nuclear function. However, the rules that oversee nuclear architecture remain unclear and controversial. As a result, there is an increasing need to replace qualitative visual assessment of microscope images with quantitative and automated methods. Such tools can substantially reduce manual labour and more importantly remove subjective bias. Quantitative methods can also increase the accuracy, sensitivity and reproducibility of data analysis. In this paper, we describe image processing and analysis methodology for the investigation of nuclear architecture, and the application of these methods to quantitatively explore the promyelocytic leukaemia (PML) nuclear bodies (NBs). PML NBs are linked with numerous nuclear functions including transcription and protein degradation. However, we know very little about the three-dimensional (3-D) architecture of PML NBs in relation to each other or within the general volume of the nucleus. Finally, we emphasise methods for the analysis of replicate images (of a given nuclear compartment across different cell nuclei) in order to aggregate information about nuclear architecture.

R.A. Russell
Department of Mathematics, Imperial College, South Kensington, London, United Kingdom
and
NIHR Biomedical Research Centre for Ophthalmology, Moorfields Eye Hospital NHS Foundation Trust and UCL Institute of Ophthalmology, London, United Kingdom

N.M. Adams
Department of Mathematics, Imperial College, South Kensington, London, United Kingdom

D. Stephens
Department of Mathematics and Statistics, McGill University, Montreal, Canada

E. Batty, K. Jensen, and P.S. Freemont (✉)
Division of Molecular Biosciences, Imperial College, South Kensington, London, United Kingdom
e-mail: p.freemont@imperial.ac.uk

N.M. Adams and P.S. Freemont (eds.), *Advances in Nuclear Architecture*,
DOI 10.1007/978-90-481-9899-3_6,

6.1 Introduction

The nucleus is the largest organelle found in eukaryotic cells. It has two major functions: first, it is the site of ribonucleic acid (RNA) synthesis; second, it stores the cell's genetic material in chromosomes and duplicates it during cell division (Alberts et al. 2002). The nucleus also contains hundreds of morphologically distinct, membrane-less protein substructures known as nuclear compartments that are responsible for the duplication, maintenance, and expression of the nucleus' genetic material. Advances in microscopy and associated fluorescence techniques have provided a wealth of nuclear image data. Such images offer the opportunity for both visualising nuclear substructures and quantitative investigation of the spatial configuration of these objects.

The complex and dynamic organisation of the cell nucleus into compartments with specific biological functions represents a crucial regulatory hub for cellular function. These compartments are tightly packed within the nucleus; at the resolution of the light microscope, two different compartments can occupy the same space. However, the spatial and temporal configuration of nuclear compartments is dynamic and responsive to changes in the nuclear environment (Rippe 2007). For an intracellular compartment as important as the nucleus, the questions that remain unanswered about its architecture are remarkably basic (Shiels et al. 2007). First, it is still unknown how nuclear compartments are distributed in the nuclear volume. A 'completely random' configuration may suggest a lack of nuclear architecture. Conversely, a 'regular' or 'aggregated' configuration may imply organisation. A second question concerns the inter-relationships between nuclear compartments. For instance, do certain compartments tend to associate, and, if so, what are the functional implications? An advanced understanding of the spatial arrangement of nuclear compartments may help to elucidate the role played by nuclear architecture in nuclear processes and help to answer whether the spatial configuration of nuclear compartments directs nuclear function, or vice-versa. This understanding should also allow a description of how nuclear architecture correlates with the functional state of the cell.

The confocal microscope has proven to be a remarkable innovation for imaging cell structure and physiology. The imaging technique provides the opportunity to visualise a specimen in three dimensions, and in time. The usefulness of the confocal microscope depends on its capacity to remove out-of-focus light, thus allowing it to capture sharp, high-contrast optical sections of cells and their sub-cellular compartments within thick samples (Dailey et al. 1999). Confocal laser scanning microscopy (CLSM) (Pawley 2006) is a technique that provides true 3-D optical resolution. In CLSM, a laser is used to form a spot that is scanned across a sample to non-invasively capture sub-micron details of cells in order to record a two-dimensional (2-D) image. This process is repeated at multiple positions in the Z-axis and image sections are then combined into an 'image stack' that yields an accurate representation of the three-dimensional (3-D) structure of the specimen. The resolution of CLSM is limited by the spot size, which depends on the wavelength of the light used and the numerical aperture of the objective and condenser lenses. Typically, the lateral resolution of the CLSM is only 0.2 μm (Pawley 2006). However, super high resolution light microscopy techniques have been developed that radically improve upon the limiting resolution of CLSM. For example, Stimulated emission depletion (STED) microscopy, is a relatively new fluorescence microscopy technique that overcomes the diffraction limit of CLSM by employing a second laser beam in the excitation spot, which selectively quenches fluorescent markers at the periphery of the spot (Hell and Wichmann 1994; Klar and Hell 1999). A resolution of up to 5.8 nm can be achieved using STED (Rittweger et al. 2009).

For now, CLSM remains the most popular technique to image nuclear architecture. In combination with an expanding array of fluorescently labelled antibodies directed against specific antigens, the device is offering novel information about the spatiotemporal dynamics of the cell and the cell nucleus; for example, see Fig. 6.1. However, it should be noted that visualisation of nuclear proteins by immunofluorescence necessitates that cells are fixed and permeabilised, which may affect the native state of the cell. Thus, cell preparation methods should be kept to a minimum (Shiels et al. 2007).

Studies employing immunofluorescent CLSM have shown that a number of nuclear compartments have preferred spatial associations but it is still unclear whether nuclei have common rules that define high-level spatial functional organisation. To address this, statistical and computational methodology is required to provide adequate quantitative image analysis tools to study and explore the subtle principles behind nuclear architecture. Human vision is superb for qualitative tasks but computers are often needed in image analysis to aid a human operator to extract quantitative information from an image (Sonka et al. 1993]. Furthermore, there is a great deal of variability in images of the cell nucleus, arising from diverse sources (but primarily innate biological variability). Therefore, we should avoid placing emphasis on individual images such as those seen in Fig. 6.1, and instead look (using quantitative image analysis methods) for aspects of spatial organisation that are common across replicate images because this truly represents underlying nuclear architecture. Quantitative image analysis is the process of making quantitative structural measurements from an image (or replicate images) via a number of distinct stages, namely image visualisation, image enhancement, segmentation, measurement and inference (Russ 1995). Importantly, specimen preparation and microscopy procedures for successful automated image analysis are generally stricter than for manual methods since computers are easily misled by artefacts, variability, confounding objects and clutter (Shiels et al. 2007). Thus, it is important to ensure that the objects of interest are delineated with a high degree of contrast against the background.

6.2 Segmentation of CLSM Image Data

A computer image can be regarded as an array of measured values. In CLSM images, this measured value is the intensity of transmitted light at a specific wavelength, and is recorded at a finite number of small geometrical subunits (referred to as pixels in 2-D images and as voxels in 3-D images). For an 8-bit image, intensity is recorded in

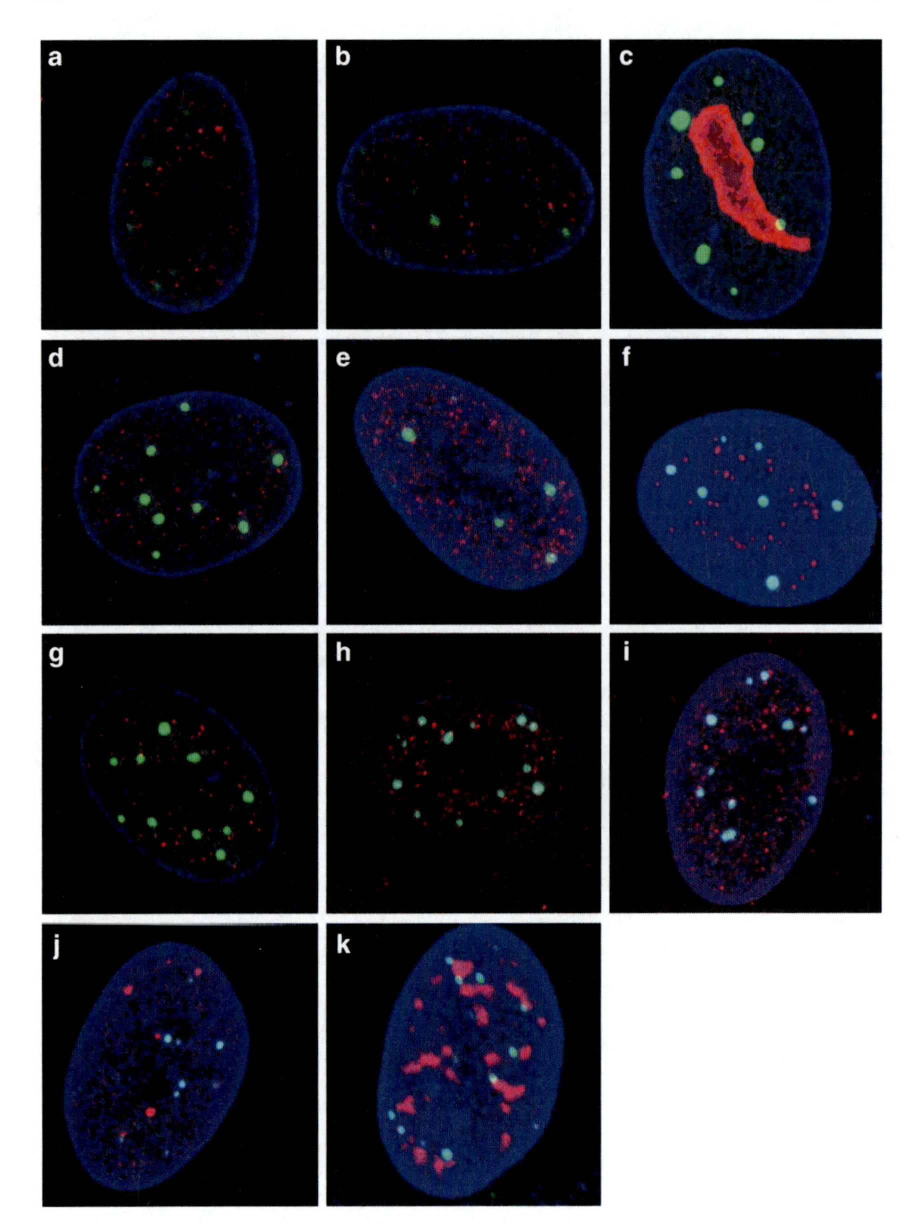

Fig. 6.1 CLSM images (shown as projections in the XY plane) of nuclear compartments in human fibroblast (MRC-5) cell nuclei. The green objects are promyelocytic leukaemia (PML) nuclear bodies (NBs), the blue object is lamin B, the red objects are: (**a**) acetylated histones; (**b**) methylated histones; (**c**) nucleoli; (**d**) RNA; (**e**) RNA polymerase II; (**f**) centromeres; (**g**) telomeres; (**h**) 11S proteasomes; (**i**) 19S proteasomes; (**j**) Cajal bodies; (**k**) SC35 domains

the range [0, 1, 2, …, 255] (images are termed as 8-bit because there are 2^8 discrete intensity values). CLSM digitally records the intensity of light in a sample at a subset of voxel locations; immunofluorescent CLSM images are multispectral since the intensity variate is measured in the RGB parts of the electromagnetic spectrum.

For quantitative analysis of nuclear architecture, image segmentation continues to represent a major first hurdle. In the analysis of image data it is essential to distinguish between the different objects of interest as well as the background. Techniques employed to distinguish the objects are referred to as segmentation methods (Sonka et al. 1993; Russ 1995). Image segmentation is a central problem in image analysis. In segmented images, parts of the image are joined into meaningful, non-overlapping regions that are believed to belong to the same objects. 'Complete segmentation' results in a set of disjoint regions uniquely corresponding to objects in the input image. Images can be segmented using a variety of approaches that can be categorised according to the general principles they employ: threshold techniques, edge-based methods and region-based techniques (Glasbey and Horgan 1995). Threshold techniques are effective when the intensities of pixels of an object are significantly different from the intensity of pixels belonging to the background (or another object). Edge-based methods rely upon contour detection. Region-based methods usually partition the image into connected regions by grouping neighbouring pixels of similar intensity levels and then merge adjacent regions according to homogeneity or sharpness of region boundaries (Glasbey and Horgan 1995). Hybrid methods are also popular and employ a mixture of the above methods.

The problem of automated image segmentation is made substantially easier when the intensities of 'object' pixels/voxels are significantly different to the intensities of the 'background' pixels/voxels (Sezgin and Sankur 2004). In such cases, thresholding is the most obvious segmentation technique to separate the object from background. Thresholding is one of the most commonly used segmentation techniques and it is also the oldest segmentation method (Sonka et al. 1993). It is a popular segmentation technique in document image analysis (to extract printed characters), quality inspection (to detect defective parts in materials) and non-destructive testing applications such as ultrasonic imaging, thermal imaging, x-ray computed tomography, CLSM and endoscopic imaging (Sezgin and Sankur 2004).

If we denote voxel location by (X, Y, Z), (the row, column and image section respectively), then thresholding segments an 8-bit input image voxel to an output binary image voxel ($S_{X,Y,Z}$) as follows:

$$\begin{aligned} S_{X,Y,Z} &= 1 \quad \text{for} \quad I_{X,Y,Z} \geq T, \\ S_{X,Y,Z} &= 0 \quad \text{for} \quad I_{X,Y,Z} < T, \end{aligned} \tag{6.1}$$

where $I_{X,Y,Z}$ is the voxel intensity with range [0, 1, 2, …, 255] , T is the threshold, $S_{X,Y,Z} = 1$ for image elements of objects, and $S_{X,Y,Z} = 0$ for image elements of the background (or vice versa) (Sonka et al. 1993). Computationally, thresholding is less demanding than most other segmentation methods and it is often easier to implement successfully.

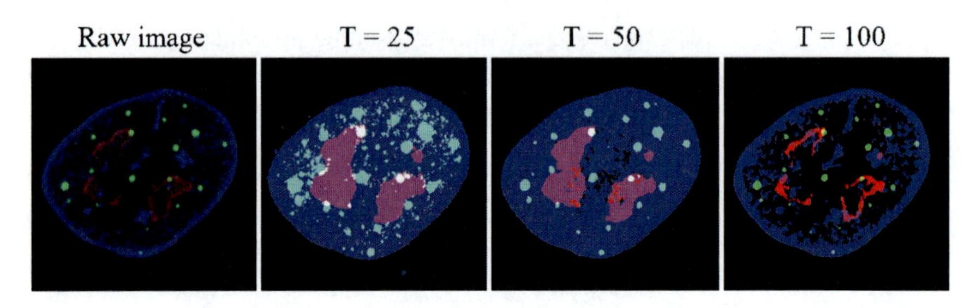

Fig. 6.2 Thresholding of PML NBs (*green*), nucleoli (*red*) and lamin B (*blue*) in a CLSM image stack (projected in the XY plane). "T" indicates the 8-bit threshold value employed

Global thresholding, as described above, is the most basic form of thresholding since it employs a single threshold value to segment an entire image. It relies on the pixels in each region having relatively homogeneous intensity; hence, global thresholding will not work well when images have an uneven background. Currently, user-defined global thresholding is a very common approach for segmenting CLSM images of the cell nucleus, e.g. (McManus et al. 2006; Shiels et al. 2001; Wang et al. 2004). The choice of this global threshold is crucial since further processing and analysis of the distinct compartments entirely depends on the quality of the segmentation. The importance of threshold selection is demonstrated in Fig. 6.2. The sizes and numbers of the segmented nuclear compartments are heavily affected by the choice of threshold. It is also very difficult, if not impossible, to determine which threshold best segments the image since we have no knowledge of the ground truth. User-defined thresholding is generally considered the gold standard for segmentation of CLSM images since the human visual system outperforms most algorithms at qualitative tasks (Glasbey and Horgan 1995). While such thresholding may be accurate it is fundamentally subjective, and this generates a demand for automated methods that perform as well as manual thresholding. Furthermore, automated methods are becoming increasingly desirable to cope with high-throughput microscopy techniques since they eliminate the time-consuming labour associated with manual thresholding.

Most automated segmentation algorithms have been designed for 2-D images. Thus, these algorithms generally segment CLSM image stacks slice by slice, losing valuable information about the 3-D image set. Some thresholding algorithms have been designed for 3-D microscopy images but their applications are limited and tend to focus on the task of cell or nucleus segmentation (Kozubek et al. 1999; Li et al. 2007; Xavier et al. 2001). We have developed a global thresholding algorithm, Stable Count Thresholding (SCT), to segment nuclear compartments in CLSM image stacks in order to facilitate quantitative analysis of the three-dimensional spatial organisation of these objects using formal statistical methods. We validated the efficacy and performance of the SCT algorithm using real images of immunofluorescently stained nuclear compartments (for example, see Fig. 6.3) and fluorescent beads as well as simulated images. In all three cases, the SCT algorithm delivered a segmentation that is far better than standard thresholding methods, and more importantly, is comparable to manual thresholding results. By applying the SCT algorithm

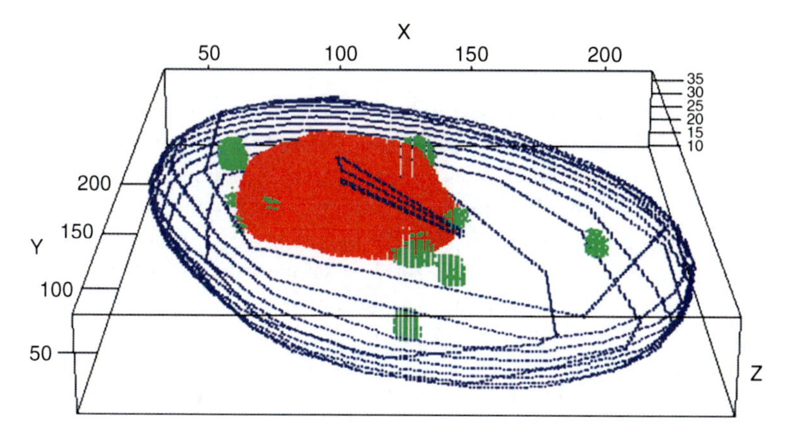

Fig. 6.3 Scatter plot of an SCT-segmented CLSM image of PML NBs (*green*), nucleolus (*red*) and lamin B (*blue*)

and statistical analysis, we were able to quantify the spatial configuration of PML NBs with respect to irregular-shaped SC35 domains. We showed that the compartments were closer than expected under a null model for their spatial point distribution, and furthermore that their spatial association varies according to cell state. More information on the SCT algorithm can be found in (Russell et al. 2009].

Image segmentation remains the heart of image analysis but, interestingly, has started to become less important due to the increasing quality of image detectors (Kozubek 2006). Recently, high-throughput microscopy has generated a need for image registration methods (Kozubek 2006). Imaging different cell nuclei (or the same cell nucleus at different times) leads to images in different coordinate systems that need to be registered; image registration is a task that we believe can be used to aggregate spatial information about nuclear architecture.

6.3 Investigating Nuclear Architecture

Having addressed the difficult intermediate problem of image segmentation, we turn our attention to quantitative assessment of segmented images. To date, the study of nuclear architecture has been dependent largely upon 'observational studies', in which immunofluorescence microscopy methods are used to image nuclear compartments and subjective visual assessment is used to analyse the images (Batty et al. 2009). This visual assessment of nuclear architecture can be inherently unreliable; human operators can often miss objects and infer different spatial relationships from the same image at different times. Thus, there is a substantive need to employ quantitative methods that take into account the statistical significance of a spatial 'relationship' between nuclear compartments, particularly given the complexity and dynamic nature of nuclear function. A variety of methods have been used to assess nuclear architecture in cell biology research, which are summarised here.

6.3.1 Radial Analysis

Radial analyses are sometimes used to assess the spatial configuration of nuclear compartments (Cremer et al. 2001; Zink et al. 2004). The centre of the nucleus is used as reference point to which the location of the nuclear compartment of interest can be calculated. Usually, location is expressed as the distance between the geometric centre of the compart-ment and the nuclear centre, normalised to remove variation in nuclear size (Shiels et al. 2007). Alternatively, the nucleus is divided into concentric shells of equal volume and the amount of fluorescent signal in each shell is used to evaluate global radial positioning. Radial analyses provide a fast means to investigate the locations of nuclear compartments between cell types or experimental conditions (Shiels et al. 2007). However, the functional implications of the radial position of a nuclear compartment are very difficult to decipher since the nuclear centre itself has no functional role. Furthermore, radial analysis offers a very limited description of spatial preference based on the distance and angle(s) between the compartment of interest and the nuclear centroid.

6.3.2 Co-Localisation Analysis

Co-localisation analysis is often employed to study whether two nuclear compartments occupy the same space (Paddock et al. 1997). The degree of co-localisation can be judged via visual examination, or using a threshold or intensity-based approach. Visual examination of the RGB image will indicate co-localisation since the overlap appears as a different additive colour. For example, overlapping green and red pixels attain a yellow appearance, which, to a first approximation indicates the presence of interacting red and green species. Visual examination of overlaid images is a relatively quick and straightforward method for detecting co-localisation between nuclear compartments, but it is strictly qualitative. 'Threshold-based' analysis can be applied in order to quantify the degree of co-localisation. This approach usually reports a 'percentage co-localisation' value that is equal to the fraction of pixels having intensity above a certain threshold. However, this approach is still subject to user-bias since a technician generally defines the threshold value. 'Intensity-based' analysis removes this bias by analysing the intensity of all the pixels in the image though some authors consider this a disadvantage due to the intrinsic uncertainty of pixel intensity, which makes comparisons between different images difficult (Lachmanovich et al. 2003). The largest disadvantage of all co-localisation analyses is that co-localisation is a very specific example of spatial organisation and is not capable of identifying more complex relationships. For example, nuclear compartments may also be found near to, but not co-localised with other compartments. In such cases, spatial association can only be quantified using distance-based methods. Naturally, the resolution of the microscope used to image objects of interest is vital for co-localisation analysis. Two objects that appear to co-localise at the limiting resolution of a microscope, may not truly overlap if the resolution were increased. Of course, this is true for all quantitative image analyses of nuclear architecture – the accuracy of any measurement will always be dependent on the resolution of the microscope.

6.3.3 Distance-Based Analysis

Distance-based methods analyse the distribution of distances measured between the nuclear compartments of interest. Typically distances are measured from the geometric centre of each compartment. Distance-based methods have been used to investigate the nuclear architecture of genomic loci (LaSalle and Lalande 1996; Neves et al. 1999), chromosome regions (Roix et al. 2003; Nikiforova et al. 2000; Quina and Parreira 2005; Bolzer et al. 2005) and PML NBs (Shiels et al. 2001; Wang et al. 2004). In Bolzer et al. (2005), inter-chromosome territories distances and Kolmogorov-Smirnov tests were used to assess the spatial distribution of 46 chromosome territories simultaneously. In Wang et al. (2004), the authors analysed the distances between PML NBs and genomic regions and correlated these with the transcriptional activity of the genomic loci to show that PML NBs associate more closely with more active genomic regions.

6.3.4 Spatial Point Pattern Analysis

Recently, a spatial point process framework has been applied to assess the spatial preference of nuclear compartments (Fleischer et al. 2006). The spatial configuration of nuclear compartments can be represented as a Spatial Point Pattern (SPP) in order to explore nuclear architecture quantitatively. An SPP is any data in the form of a set of points, distributed within a region of space (Diggle 2003). Examples of such data include locations of taste buds on a tongue, trees in a forest, or stars in the sky (Bell and Grunwald 2004). Locations of objects are referred to as 'events' in order to differentiate them from other points in the space. Processes that generate SPPs can be broadly categorised as producing events that are: completely random (events lie uniformly in the region and independently of each other), aggregated (events are clustered together) or regular (events are arranged in a periodic fashion). The simplest spatial point process model is a homogeneous spatial Poisson process, which exhibits what is known as complete spatial randomness (CSR). Under CSR, events are distributed independently and are equally likely to occupy any part of the region.

SPP analysis often involves comparisons between empirical summary descriptions of distance data and the corresponding theoretical summary descriptions of a spatial point process model (Diggle 2003). We distinguish 'distance-based' analysis from SPP analysis since the former does not require a theoretical summary description (derived from an underlying model) of the distance data. 'Distance-based' analyses are generally concerned with comparing distances from two different samples. For example, in (Wang et al. 2004), the authors showed that the distances between PML NBs and 'active' genomic regions were smaller than those for 'inactive' genomic regions (Wang et al. 2004). If a point pattern is shown to exhibit CSR in some bounded region then it is a realisation of a homogeneous spatial Poisson process. This implies that the locations of the events are distributed uniformly and independently of one another within the region. If CSR is rejected then the extent and direction of the departure can guide the choice of alternative models to be formulated for the SPP. More background on spatial point pattern analysis is given in many textbooks including (Diggle 2003) and (Illian et al. 2008).

In Russell et al. (2009), we employed SPP analysis in order to explore nuclear architecture. We probed the 3-D spatial interactions of SCT-segmented PML NBs with SC35 domains, within a spatial point process framework. The PML NB, a nuclear compartment linked with numerous nuclear functions including transcription and protein degradation may locate in close proximity to, adjacent to, or even co-localise with other functional nuclear compartments, such as Cajal bodies, SC35 domains, and telomeric DNA, respectively (Bernardi and Pandolfi 2007; Borden 2002; Cressie 1993). Our analysis employed a summary statistic, known as the G-function, which is the empirical distribution function (EDF) of nearest neighbour distances (NNDs) between the objects of interest, known as 'events'. The G-function is a standard tool for exploratory analysis of SPP data. Our approach compared the observed distance distribution against a null distribution estimated by Monte Carlo simulation in the given nucleus (Russell et al. 2009; Diggle 2003). Our Monte Carlo simulation procedure was enhanced to take into account both volume and shape variability of the given nucleus as well as variability in the number and volume of the given PML NBs. To assess whether any given SPP rejects the null hypothesis of CSR, we examine the observed G-function using 999 simulated realisations to define upper and lower limits of a 99% simulation envelope. To illustrate this analysis, for a specific cell nucleus (shown in Fig. 6.3), we have investigated the SPP of PML NBs with respect to the nucleolus; see Fig. 6.4. In Fig. 6.4, the G-function is illustrated by the red line and the 99% simulation envelope is shaded

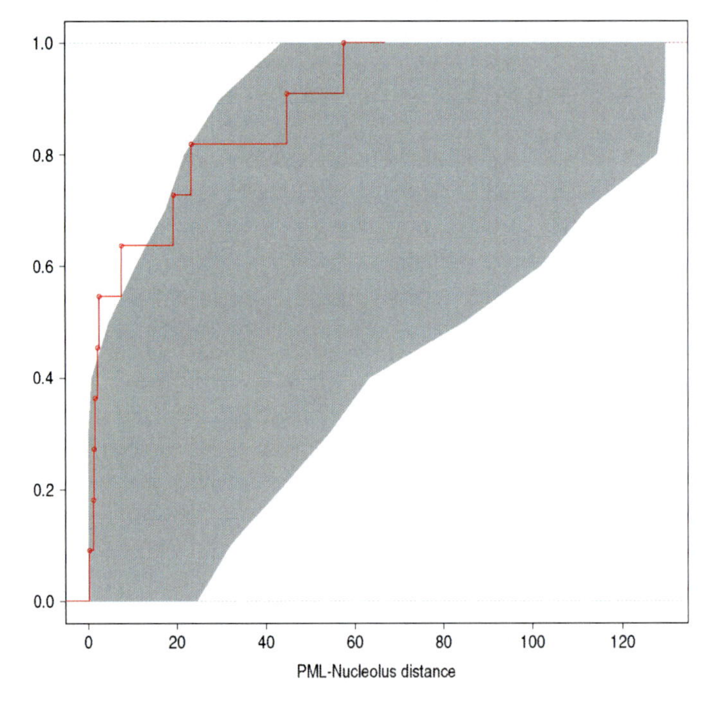

Fig. 6.4 Plot of the G-function of observed NNDs between PML NBs and the nucleolus shown in Fig. 6.3. The 99% simulation envelope is shaded in gray

in gray. The figure suggests that this SPP is not consistent with the null hypothesis since the G-function is not fully contained by the simulation envelope.

It is worth emphasising that no matter what type of analysis is used to assess nuclear architecture, the interpretation of the findings is difficult. For example, if nuclear compartments are shown to have a statistically significant association, further biochemical studies are essential to translate such observations into a functional interaction. Our application of SPP methodology in Russell et al. (2009) was wholly an exploratory analysis of PML NBs. The most significant limitation of our analysis was that – for the most part – we treated the data as 'un-replicated' (Everett 2001) i.e., we explored each SPP in each nucleus separately. Pooling information across all nuclei is a hard problem since nuclear image data consists of SPPs in regions of variable size and shape. However, image registration methodology offers the means to aggregate spatial information about nuclear architecture across image data of different cell nuclei.

6.4 The Future of Analysing Nuclear Architecture – Image Registration?

To date, most analyses of nuclear architecture have given little insight into the underlying spatial distribution of nuclear compartments in the cell nucleus. We believe that image registration methodology can be used in order to combine spatial information across multiple cell nuclei. Image registration is the process of bringing two or more images into spatial alignment. There are two generic approaches to image registration: intensity-based and feature-based (Diggle et al. 2006; Friston et al. 2006; Rohr 2001). Intensity-based approaches operate directly on image intensities and therefore an explicit segmentation of the images is not required. In contrast, feature-based methods involve specific processing of the raw images to identify physical homologous features. Generally, intensity based methods are only used when feature-based methods are unavailable since the former are sensitive to image noise and variations in the image such as lighting conditions (Friston et al. 2006). Recently, an intensity-based image registration method has been applied to fluorescent CLSM images of cell nuclei (Zitova and Flusser 2003). In this research, the authors investigate the 3-D structure of the X-chromosome. They argue that feature-based approaches are difficult because identification of homologous features usually requires some kind of user-interaction. Since intensity-based approaches are sensitive to image noise, our current research investigates the question, "is it possible to construct an automatic feature-based image registration procedure to combine images of cell nuclei?"

In order to combine information across images, we need to reason about nuclear shape. If cell nuclei were of the same shape and size then combining information on nuclear architecture would be an easy task – spatial information on nuclear compartments could be very quickly combined after removing differences in the orientation and location of imaged nuclei. However, cell nuclei vary substantially in shape and size. To aggregate information, we wish to combine our raw data such that the imaged nuclei are 'normalised'; i.e., transformed to have the same shape and size. The alternative approach is to analyse the nucleus in each image stack separately and then aggregate this information for all nuclei. We are interested in the former approach; however, this task is made difficult since no landmarks have been identified to allow the spatial normalisation of nuclear shape and size. A possibility is to use the nuclear envelope as a landmark, with which replicate nuclei can be put in a common coordinate system. Both cells and nuclei have an envelope which fundamentally delimit them and undergoes significant changes in size and shape through the cell cycle. Our interest in these envelopes is threefold. First, learning about the shape of the envelope is of interest in its own right. For example, if subtle shape differences can be associated with different parts of the cell cycle, automated classification methods can be used to identify cells, avoiding the need for difficult experimental procedures that explicitly control cell cycle. Second, the envelope provides the only available landmark. As noted earlier, the popularity of radial distance tests is striking. The centroid of the cell nucleus has no special meaning, so its use as a landmark is questionable at best. In contrast, the envelope is a genuine object that is only occasionally used in spatial comparisons. We have already discovered interesting relationships between PML bodies and the envelope using SPP methodology. Finally, and most ambitiously we can consider registering replicate cell nuclei according to their envelope, such that they are placed in a common coordinate system. This opens the possibility to reason about the *average* configuration of a nuclear compartment across replicate images.

Figure 6.5 represents an initial attempt to construct such an average representation from a set of 50 replicate images of fibroblast (MRC-5) cell nuclei. The figure illustrates the 2-D projection of the average-shaped nuclear envelope with a smoothed estimate of the density of SC35 domains contained within the collection of registered nuclei. We begin with a number of prepared cell images, similar to Fig. 6.1, that have been segmented using the SCT algorithm. Each segmented image is then represented as a 2-D pixel map. Subsequently, the sequence of steps required to produce Fig. 6.5 are

1. Identity a common set of mathematical pseudo-landmarks for *each* nuclear envelope
2. Register each nuclear envelope to the average envelope, on the basis of its landmarks
3. Transform the SC35 domains from each image into the average envelope, in a manner that respects the registration of each envelope

Figure 6.5 demonstrates a clear spatial preference for SC35 domains, in this case, to prefer a central region of the nucleus. Certain aspects of this have been revealed by the more traditional approaches described in the previous section – but the strength and clarity of the signal here is unprecedented. Note also that this representation refers to *normal* cells, and provides another means of clarifying the effects of experimental perturbation. We have striking results for a range of nuclear compartments in MRC-5 cells, including experimental perturbations, related to heat shock. These appear to be hinting at fundamental ideas of spatial nuclear organisation.

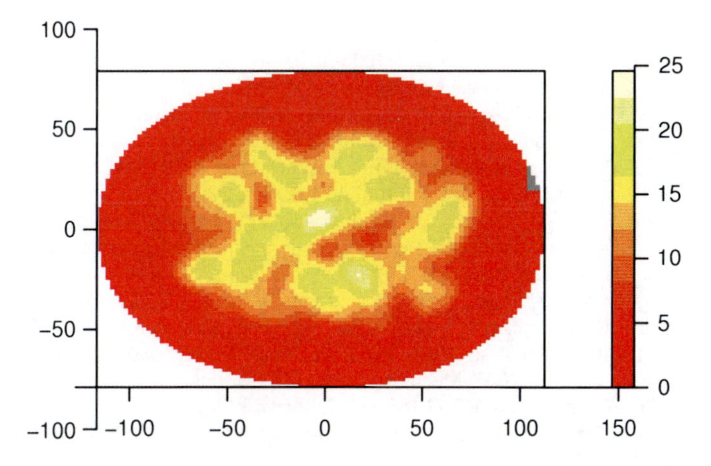

Fig. 6.5 Collection of registered MRC-5 cell nuclei, with SC35 domains spatial density estimates

6.5 Conclusion

This paper has presented image processing techniques as well as statistical and computational tools to explore and uncover the rules that oversee nuclear architecture in biological confocal microscopy image data. Such research combines a series of stages from image processing to image analysis. As such, we have addressed each stage of this process in order to have a consistent and objective approach to the exploration of nuclear architecture.

Our current research aims to develop a procedure that provides a spatial normalisation of replicate cell nuclei image data by combining image registration and statistical shape analysis methodology in a novel way. Current image analysis tools are generally based on exploring spatial organisation in individual images. Investigation of nuclear architecture from replicate images of cell nuclei probes absolute spatial organisation that cannot immediately be established from analysis of individual images. In this way, spatial normalisation provides us with a 'virtual cell nucleus' such as Fig. 6.5. Strikingly, this virtual nucleus provides new and exciting insights about high-level nuclear architecture, suggesting that nuclear compartments preferentially occupy distinct regions of the nucleus in 'normal' cell nuclei.

References

Alberts B, Johnson A, Lewis J, Raff M, Roberts K, Walter P (2002) Molecular biology of the cell, 4th edn. Garland, New York

Batty E, Jensen K, Freemont PS (2009) PML nuclear bodies and their spatial relationships in the mammalian cell nucleus. Front Biosci 14:1182–1196

Bell ML, Grunwald GK (2004) Mixed models for the analysis of replicated spatial point patterns. Biostatistics 5(4):633–648

Bernardi R, Pandolfi PP (2007) Structure, dynamics and functions of promyelocytic leukaemia nuclear bodies. Nat Rev Mol Cell Biol 8:1006–1016

Bolzer A, Kreth G, Solovei I, Koehler D, Saracoglu K, Fauth C, Müller S, Eils R, Cremer C, Speicher MR, Cremer T (2005) Three-dimensional maps of all chromosomes in human male fibroblast nuclei and prometaphase rosettes. PLoS Biol 3:e157

Borden KL (2002) Pondering the Promyelocytic Leukemia Protein (PML) puzzle: possible functions for PML nuclear bodies. Mol Cell Biol 22(15):5259–5269

Cremer M, von Hase J, Volm T, Brero A, Kreth G, Walter J, Fischer C, Solovei I, Cremer C, Cremer T (2001) Non-random radial higher-order chromatin arrangements in nuclei of diploid human cells. Chromosome Res 9:541–567

Dailey M, Marrs G, Satz J, Waite M (1999) Concepts in imaging and microscopy exploring biological structure and function with confocal microscopy. Biol Bull 197(2):115–122

Diggle PJ (2003) Statistical analysis of spatial point patterns, 2nd edn. Arnold, London

Diggle SJ, Eglen PJ, Troy JB (2006) Modelling the bivariate spatial distribution of amacrine cells. Case Stud Spat Point Process Model 185:215–233

Everett RD (2001) DNA viruses and viral proteins that interact with PML nuclear bodies. Oncogene 20(49):7266–7273

Fleischer F, Beil M, Kazda M, Schmidt V (2006) Analysis of spatial point patterns in microscopic and macroscopic biological image data. Case Stud Spat Point Process Model 185:235–260

Friston KJ, Ashburner JT, Kiebel SJ, Nichols TE, Penny WD (2006) Statistical parametric mapping: the analysis of functional brain images. Academic, London

Glasbey CA, Horgan GW (1995) Image analysis for the biological sciences. Wiley, New York

Hell SW, Wichmann W (1994) Breaking the diffraction resolution limit by stimulated emission: stimulated-emission-depletion fluorescence microscopy. Opt Lett 19(11):780–782

Illian J, Penttinen A, Stoyan H, Stoyan D (2008) Statistical analysis and modelling of spatial point patterns. Wiley, New York

Klar TA, Hell SW (1999) Subdiffraction resolution in far-field fluorescence microscopy. Opt Lett 24(14):954–956

Kozubek M (2006) Automated analysis of multi-dimensional biomedical image data acquired using optical microscopy. Proceedings of the 6th international conference on stereology, Spatial statistics and stochastic geometry, pp 465–476

Kozubek M, Kozubek S, Lukásová E, Marecková A, Bártová E, Skalníková M, Jergová A (1999) High-resolution cytometry of FISH dots in interphase cell nuclei. Cytometry 36:279–293

Lachmanovich R, Shvartsman DE, Malka Y, Botvin C, Henis YI, Weiss AM (2003) Co-localization analysis of complex formation among membrane proteins by computerized fluorescence microscopy: application to immunofluorescence co-patching studies. J Microsc 212(2):122–131

LaSalle JM, Lalande M (1996) Homologous association of oppositely imprinted chromosomal domains. Science 272:725–728

Li G, Liu T, Tarokh A, Nie J, Guo L, Mara A, Holley S, Wong ST (2007) 3D cell nuclei segmentation based on gradient flow tracking. BMC Cell Biol 8(1):401

McManus KJ, Stephens DA, Adams NM, Islam SA, Freemont PS, Hendzel MJ (2006) The transcriptional regulator CBP has defined spatial associations within interphase nuclei. PLoS Comput Biol 2(10):e139

Neves H, Ramos C, da Silva MG, Parreira A, Parreira L (1999) The nuclear topography of ABL, BCR, PML, and RARalpha genes: evidence for gene proximity in specific phases of the cell cycle and stages of hematopoietic differentiation. Blood 93:1197–1207

Nikiforova MN, Stringer JR, Blough R, Medvedovic M, Fagin JA, Nikiforov YE (2000) Proximity of chromosomal loci that participate in radiation-induced rearrangements in human cells. Science 290:138–141

Paddock SW, Hazen EJ, DeVries PJ (1997) Methods and applications of three colour confocal imaging. BioTech 22:120–126

Pawley JB (2006) Handbook of biological confocal microscopy, 3rd edn. Plenum, New York

Quina AS, Parreira L (2005) Telomere-surrounding regions are transcription-permissive 3D nuclear compartments in human cells. Exp Cell Res 307:52–64

Rippe K (2007) Dynamic organization of the cell nucleus. Curr Opin Genet Develop 17:373–380
Rittweger E, Han KY, Irvine SE, Eggeling C, Hell SW (2009) STED microscopy reveals crystal colour centres with nanometric resolution. Nat Photon 3:144–147
Rohr K (2001) Landmark-based image analysis: using geometric and intensity models. Kluwer, London
Roix JJ, McQueen PG, Munson PJ, Parada LA, Misteli T (2003) Spatial proximity of translocation-prone gene loci in human lymphomas. Nat Genet 34(3):287–291
Russ JC (1995) The image processing handbook, 2nd edn. CRC, Boca Raton
Russell RA, Adams NM, Stephens DA, Batty E, Jensen K, Freemont PS (2009) Segmentation of fluorescence microscopy images for quantitative analysis of cell nuclear architecture. Biophys J 96(8):3379–3389
Sezgin M, Sankur B (2004) Survey over image thresholding techniques and quantitative performance evaluation. J Elec Imag 13(1):146–165
Shiels C, Adams NM, Islam SA, Stephens DA, Freemont PS (2007) Quantitative analysis of cell nucleus organisation. PLoS Comput Biol 3(7):e138
Shiels C, Islam SA, Vatcheva R, Sasieni P, Sternberg MJE, Freemont PS, Sheer D (2001) PML bodies associate specifically with the MHC gene cluster in interphase nuclei. J Cell Sci 114:3705–3716
Sonka M, Hlavac V, Boyle R (1993) Image processing, analysis and machine vision. International Thomson Computer Press, London
Wang J, Shiels C, Sasieni P, Wu PJ, Islam SA, Freemont PS, Sheer D (2004) Promyelocytic leukemia nuclear bodies associate with transcriptionally active genomic regions. J Cell Biol 164:515–526
Xavier JB, Schnell A, Wuertzb S, Palmer R, White DC, Almeida JS (2001) Objective threshold selection procedure (OTS) for segmentation of scanning laser confocal microscope images. J Microbiol Meth 47(2):169–180
Yang S, Illner D, Teller K, Solovei I, van Driel R, Joffe B, Cremer T, Eils R, Rohr K (2008) Structural analysis of interphase X-chromatin based on statistical shape theory. Biochimica Biophysica Acta 1783(11):2089–2099
Zink D, Fischer AH, Nickerson JA (2004) Nuclear structure in cancer cells. Nat Rev Cancer 4:677–687
Zitova B, Flusser J (2003) Image registration methods: a survey. Image Vision Comput 21: 977–1000

Chapter 7
Thinking Holistically About Gene Transcription

Dean A. Jackson

Abstract Gene expression in higher eukaryotes demands a highly orchestrated series of events that is regulated at many levels. This hierarchical control begins in the nucleus where gene expression is activated by gene transcription. The control of transcription itself is multi-layered and incorporates both genetic and epigenetic features. Epigenetic regulation involves post-translational modification of histones and other chromatin proteins, which define the local chromatin environment of a gene and organizational features, which define the nuclear environment. In this essay, I explore how the nuclear environment can contribute to the regulation of gene expression. I discuss very recent experiments that provide compelling evidence for the widespread formation of gene expression networks during the induction of gene expression and evaluate how the dynamic behaviour of chromatin, which is required during the formation of such networks, fits with present models of nuclear organization.

Keywords Gene expression • Systems biology • Nuclear architecture • Nuclear organization • Nuclear compartments • Nuclear matrix • Nucleoskeleton • RNA transcription • Transcription factory • Gene networks • Chromosome structure • DNA foci • Modelling gene expression • NF-κB

7.1 Introduction

Over the past 30 years, biological research has undergone a series of dramatic shifts in emphasis. Working on the structure and function of mammalian nuclei over this time, it is remarkable how our experimental abilities have evolved, even if the same fundament questions remain unresolved. The 1980s saw the heyday of

D.A. Jackson (✉)
University of Manchester, Faculty of Life Sciences, 131 Princess St, Manchester, M1 7DN, UK
e-mail: dean.jackson@manchester.ac.uk

N.M. Adams and P.S. Freemont (eds.), *Advances in Nuclear Architecture*,
DOI 10.1007/978-90-481-9899-3_7,

molecular biology, as reductionist strategies of the 1970s gave biochemists of the impetus to understand how interactions between molecules might be controlled to regulate key aspects of cell function. In the 1990s, the development of fluorescent protein technologies and later RNA interference took the molecular focus towards a cellular context. More recently, the emphasis to think more holistically about biological systems has heralded the growth of modelling driven analyses, under the umbrella of 'systems biology'.

Systems biology aims to approach biological processes from a holistic perspective. In the nucleus, this strategy to addressing key biological questions clearly represents a paradigm shift. However, the approach is not without precedent, and as we continue to look deeper into the mysteries of nuclear structure and function the molecular complexity ensures that it is no longer sensible to consider individual components in isolation. As an example, it is interesting to consider the biomedical implications of understanding circadian rhythms (Ko and Takahashi 2006). The mammalian circadian clock couples physiological functions to environmental cues, most significantly light and temperature (Matsuo et al. 2003). The timing system is controlled by the suprachiasmatic nuclei within the hypothalamus and involves a complex network of clock genes and regulatory feedback loops (Filipski and Levi 2009). The clock genes, in turn, regulate a series of fundamental biological processes including cell cycle, cell proliferation, differentiation, genome stability and apoptosis. In humans, an understanding of how these processes are controlled throughout our daily physiological cycles is becoming every more important in medicine, where treatment regimes for major diseases such as cancer and cardiovascular conditions are significantly more effective if coupled to the circadian physiology (Filipski and Lévi 2009; Wood et al. 2009; Paschos et al. 2010).

7.2 The Nucleus

As mammalian cells respond to environmental cues by regulating patterns of gene expression it is important to understand how different features of chromatin structure and global nuclear organization contribute to gene regulation (Baxter et al. 2002; Misteli 2007). The control of gene expression is by necessity complex and as a result is regulated at many different points. All of the regulatory functions are, however, dependent on the essential first steps that occur at gene promoters in order to activate the process of RNA synthesis (Emerson 2002; Maniatis and Reed 2002). Fundamental features of this early phase are defined genetically as they rely on the interaction of activating transcription factors with cognate recognition motifs in gene promoters. Initial interactions at promoters set in place a chain of events that dynamically recruits a complex array of proteins that ultimately engage the synthetic RNA polymerase complex (Hager et al. 2009). Events at the promoter also ensure that the local chromatin environment is primed to facilitate the dynamic topological changes that are required during transcription (Berger 2007;

Bernstein et al. 2007). The molecular events at promoters have been studied in great detail over recent years and though our molecular understanding continues to develop in detail the basic principles are generally agreed. However, how other aspects of nuclear organization contribute to gene expression continue to attract a significant amount of debate. In this essay I will consider if/how such features contribute to gene expression and how knowledge of any contribution might be incorporated into a modelling framework.

The following areas have been implicated in gene expression over recent years and are worthy of consideration here:

1. Implication of spatial nuclear organization related to the structure of active centres of RNA synthesis – transcription factories
2. Higher-order gene networks, how they form and if/how they contribute to the regulation of gene expression
3. Defining the possible roles of higher-order chromatin folding and both short and long-range chromatin dynamics during gene expression
4. Understanding how the architecture of nuclear compartments might facilitate or restrict the formation of functional gene networks

7.3 Transcription Factories

The development of techniques to label sites of nascent pre-mRNA synthesis in human cells provided the first evidence that transcription might be arranged within dedicated nuclear sites that contain multiple synthetic complexes (Jackson et al. 1993). Further studies confirmed that these so called 'transcription factories' contained about ten active synthetic complexes (Jackson et al. 1998) and could be visualized in fixed cells as discrete structure that were ~80 nm in diameter and contained protein complexes with a mass in excess of 10 MDa (Eskiw et al. 2008). Very recent studies have extended the factory concept by demonstrating that factories also exist with polymerase complexes that represent a pre-elongation state, implying that factories are not defined by the synthetic process itself and that during the activation of gene expression genes must be dynamic locally in order to engage local transcription machinery within pre-assembled factories (Ferrai et al. 2010).

Recent advances in our understanding of nuclear organization and gene expression suggest that the concept of transcription factories will have fundamental implications for gene expression control (Cook 1999, 2010). In fact, the events that are required to generate a mature mRNA are so highly orchestrated that the development of a structure to regulate the behaviour of the synthetic machinery seems almost inevitable. During a single transcript cycle the polymerase structure itself is central to this role, as the C-terminal domain of the largest subunit of RNA polymerase II coordinates processing of the nascent transcripts through a series of interactions that recruit the relevant processing complexes to the active site.

The fact that all of the activities required to generate mature mRNAs occur at the transcription centre support the general theme of synthetic factories. At this level of complexity, factories facilitate the temporally programmed recruitment of the processing complexes that together generate mRNA.

7.4 Gene Networks

Other lines of evidence suggest that factories have a greater purpose, as they provide active sites where genetically unrelated gene can be transcribed together and in principle co-regulated. Models describing the co-transcription of genes within a common active site have evolved over recent years (Fraser and Bickmore 2007) following seminal experiments that were designed to understand how remote regulatory sequences within the β-globin locus might contribute to the regulation of globin gene expression (Patrinos et al. 2004; Noordermeer and de Laat 2008). In the human β-globin gene cluster, individual gene promoters were shown to interact spatially, as a result of chromatin looping, so that promoters, the relevant enhancers and sequences within the distal locus control region would all interact together within a common regulator complex, which was called the active chromatin hub (ACH). During formation of the ACH, the spatial co-association of genetically linked sequence elements requires that the intervening chromatin is displaced from the hub as chromatin loops; this model provides an excellent paradigm to explain how chromatin loops form as a result of DNA interactions within protein complexes that arise during gene expression.

Experiments that were designed to investigate the structure of protein complexes such as the ACH inevitably required detailed analysis of interactions between local and remote sequence elements. However, a really surprising extension of this work showed that different genomic sites might be co-associated during gene expression even when these were not linked by there genetic location on individual chromosomes (Fig. 7.1; Osborne et al. 2004; Fraser and Bickmore 2007). In experiments described to date, inter-chromosomal gene interactions have been described in many situations, and while the co-association of the gene pairs studied has never approach 100% the observed levels of interaction far exceed those that would be expected by random chance. In fact, for the majority of examples studied the levels of inter-chromosomal interactions shows that some 5–10% of the active genes will typically co-localize at a common active site (Fig. 7.1; Schoenfelder et al. 2010); in this experimental system, a maximum co-localization of ~25% is seen for selected gene pairs (Osborne et al. 2004, 2007).

While the recognition of inter-chromosomal gene interactions clearly defines an important concept in nuclear organization the stochastic nature of these interactions implies that they are not obligatory for gene expression. However, it is actually technically challenging to provide a detailed analysis of the effects of gene interactions within common active sites when the analytical tools are only able to define these interactions at a fixed point in time. It may, for example, be argued that even though a restricted number of genes undergo productive co-localization at active sites these might increase expression if the genes within these sites produce transcripts with higher efficiency than genes that are not spatially associated with genes from the same regulatory network One example in the literature does

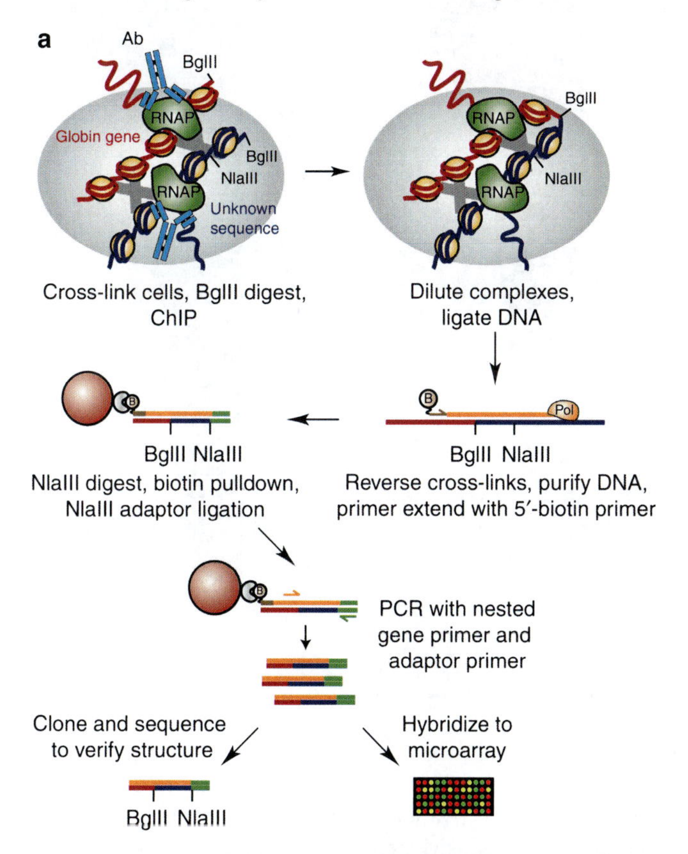

Fig. 7.1 Analysis of gene networks in nuclei of mammalian cells. The spatial architecture of genetic elements within nuclei can be probed using chromosome, conformation capture (3C) techniques (**a**). The principle behind 3C is that DNA that interacts within a molecular super-structure can be identified if the structures are first cross-linked in situ and DNA then fragmented to leave short restriction fragments in place. After sonication, to release the nuclear structures, DNA molecules that are related by their co-association within individual structures can be identified by ligating them together if the structures are first massively diluted to limit non-specific interactions. The ligated fragments can then be purified, see above, and PCR used to amplify fragments associated with selected target sequences and analysed on micro-arrays or using high volume sequence analysis. In this experiments from Schoenfelder et al. (2010) an enhanced ChIP-4C (e4C) approach was used that incorporates immunopurification of structures containing the elongation competent form of RNA polymerase II and a pre-enrichment step for bait-linked sequences by immunopurification of the biotin containing fragments. Interactions between specific gene loci must be validated in single cells using fluorescent in situ hybridization (**b**). The analysis of mouse genes that associate within the globin gene networks were assessed by their level of co-localization with *Hba* and *Hbb* (**b**). The extent of co-localization with various genes is shown (histograms, with interacting genes identified on the *left*) (Published from Schoenfelder et al. (2010) with permission of Nature Publishing Group)

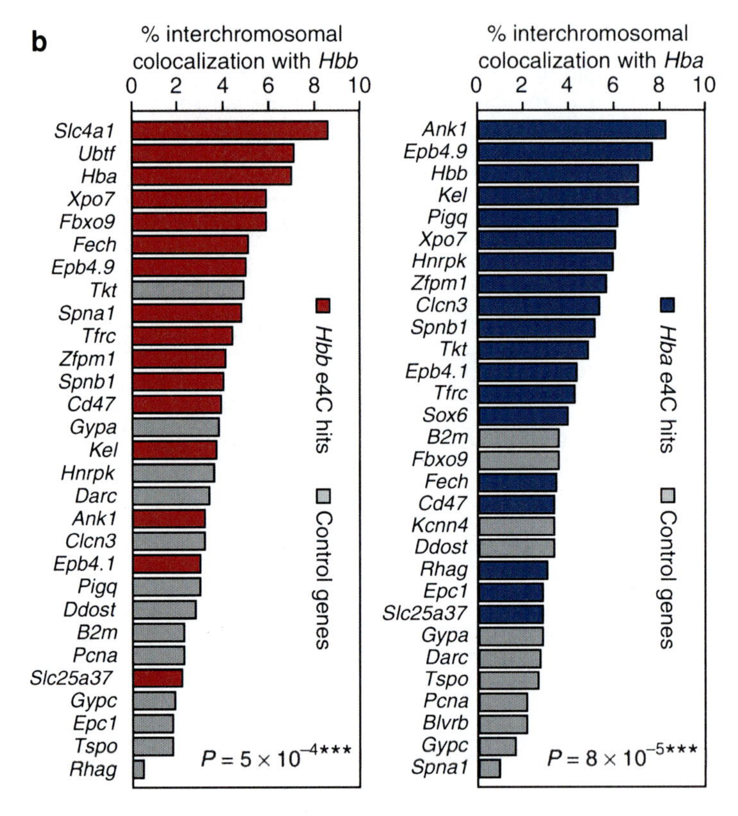

Fig. 7.1 (continued)

demonstrate that genes transcribed from a common transcription site are able to interact synergistically to provide increased levels of expression (Hu et al. 2008). This study explored the activation of hormone-induced (17β-estradiol) gene expression in oestrogen responsive genes of the transformed human breast cell line MCF7. Focussing on the *TFF1* locus on human chromosome 21, a 3C-based approach was combined with hybridization-based fragment purification and microarray analysis to define intra-chromosomal interactions as well as inter-chromosomal interactions within the regulatory elements of the *GREB1* gene on chromosome 2. After inducing gene expression for 1 h, about 50% of nuclei contained sites where *TFF1* and *GREB1* were colocalized; some cells showed bi-allelic colocalization and very few showed colocalization prior to induction. Perhaps remarkably, chromosome painting showed that in some cells, the induced inter-chromosomal interactions also correlated with relocation of the associated chromosome territory (CT; associations increased by fivefold relative to control) though this was much less than the individual interactions, suggesting that chromatin looping is more common. Using RNAi, nuclear actin, myosin and the motor protein dynein light chain-1 were all shown to be required for gene repositioning, though these treatments did not alter recruitment of transcription factors to the target promoters.

Finally, the monoallelic interactions were shown to produce a ~10-fold increase in transcription relative to the levels of synthesis at corresponding isolated loci.

7.5 Organization in the Chromatin Compartment

An obvious implication of inter-chromosomal interactions and the potential to form functional gene networks is that the genomes in question must display a high degree of spatial plasticity. However, the extent of this plasticity and more particularly the distances involved appear to contradict many studies that have show the mammalian genome to be generally rather stable for most of the cell cycle (Cremer and Cremer 2001; Lanctôt et al. 2007). Detailed studies have shown that while the fundamental structural subunits of chromosomes – so called DNA foci that contain roughly 1 Mbp of DNA on average – are locally dynamic at a range of microns, this property of chromatin is rarely seen to lead to a significant shift in the location of CTs, which in most instances establish a preferred stable location throughout interphase. Unusually, highly active gene clusters are seen to extend as chromatin loops that at their extremities might be many microns away from their associated chromosome territory (Volpi et al. 2000), though these structures are seen in only a minority of cells and their dynamic behaviour and any biological significance remains to be defined. In this context, it is important to remember that techniques that have explored global chromatin dynamics using DNA-binding dyes or fluorescent chromatin-associated proteins show that DNA is dynamic in the distance range 0–1 μm but over lager distances the general spatial architecture of chromatin is rather static over many hours (Cremer and Cremer 2001).

Our present understanding of chromatin dynamics is inevitably complicated by apparent differences in the dynamic behaviour of global DNA and specific gene domains. A possible implication of this is that the dynamic behaviour of individual domains is easily masked when the generally immobile bulk chromatin is analysed. Importantly, as it is difficult to perceive a specific mechanism that is able to 'direct' chromatin movement through the nucleoplasm and towards a specific remote target sites the most likely mechanism must involve chromatin domains that are able to probe the local inter-chromatin space in order to continually sample the range of local microenvironments. If, during this process, gene promoters within an exposed chromatin domain encounter an environment that is permissive or optimal for expression the genes will be assimilated locally into the active site where other genes with similar demands for transcription factors might be found. How is this able to happen? The fact that it happens at all implies a functional mechanism – these structures are far too big to operate by diffusion and are blocked if nuclear motor functions are inhibited (Dundr et al. 2007; Hu et al. 2008). Even if only specific classes of genes that function within wide-scale gene networks are continually able to probe the local environment to find the most productive active sites the chromosome landscape will assume a preferred steady state organization that must reflect some optimal state of productive interaction.

Mechanistically, there is some residual debate about mechanisms that result in gene association during transcription. In human cells, there is some evidence that co-associated genes are located at common nuclear speckles and that within these sites only a minority of genes interact with common transcription factories (Brown et al. 2008; Hu et al. 2008). However, more recent studies in mouse have shown that co-association occurs within factories and is dependent on specific transcription factors – the factor Klf1 (Kruppel like factor-1) provides a compelling test case – that are required to activate expression of the associated genes (Schoenfelder et al. 2010). These apparent differences may be cell type specific but may also reflect the formation of intermediate steady state structures where the transcription factories define the primary sites of co-association and speckles provide a local domain where secondary meta-stable interactions might persist. Recent evidence for the co-association of genes within individual transcription factories appears to be decisive (Papantonis et al. 2010). This study looked at genes that were induced by treating human umbilical vein endothelial cells with the cytokine tumour necrosis factor (TNF). This factor induces expression of around 300 target genes that lie down-stream of the NF-κB signal transduction pathway. Two genes were chosen for analysis. The first, *SAMD4A*, was chosen because this long gene (with a 221 kbp transcript) was shown recently to be transcribed by elongating polymerase complexes that pass along the gene as a synchronised wave, such that for each gene copy only one elongating complex is present at a time (Wada et al. 2009). With an average transcription speed of ~3 kbp/min, this gene is engaged in transcription for at least 60 min. The second gene, *TNFAIP2*, which is separated from *SAMD4A* by ~50 Mbp, was selected because this 11 kbp gene is transcribed in ~3 min. Using the 3C approach to probe co-association of these genes, this study showed that a 3C-based interaction between the genes could be mapped to pass down the SAMD4A gene as a wave that correlated precisely with the location of the elongating polymerase complex on that gene. Clearly this is only possible if the genes in question are held in close spatial proximity through their interactions with RNA polymerase within a common active site (Papantonis et al. 2010).

In coming years there is no doubt that unbiased genome-wide tools will be used to probe interaction domains within mammalian genomes at ever greater resolution. The potential power of this type of approach has been exemplified by a recent study that used a 'Hi-C' strategy to develop a genome-wide interact map (Lieberman-Aiden et al. 2009). Interactions were performed using the general 3C principle adapted for unbiased analysis with paired-end (Solexa) ultra-high volume sequencing. Analysis of the interaction networks confirmed the general separation of chromatin into two broad classes, which correlate with euchromatin and heterochromatin, and the localization of chromosome specific elements, consistent with the accepted models of chromosome distribution into discrete nuclear domains or territories. These observations provide important experimental validation of the basic strategy. Analysis of the interaction map at higher resolution revealed a general tendency for the formation of higher-order chromatin domains at a resolution of ~1 Mbp. These presumably correspond with DNA foci, which have been known for many years to correspond with higher order domains of both nuclear structure and function

(Jackson and Pombo 1998; Cremer and Cremer 2001). Interestingly, while DNA foci have been shown to be stable structure there is as yet no robust analysis of domain boundaries. Perhaps the Hi-C strategy with a higher density of sequence reads would provide important insight into the location and stability of higher-order chromatin domains in mammalian cells.

Unbiased multi-C approaches hint at the complexity of interactions that might be engaged by a specific gene locus. However, it is important to understand that these experiments will generally involve huge cell populations (typically $>10^6$) so provide information about the possibility of interaction rather than details about the frequency of interactions within individual cells. Analysis of the efficiency of interaction is only possible if contacts between putative interaction partners can be visualized within individual cells. Fluorescent in situ hybridization (FISH) techniques allow interactions between specific domains to be mapped, and many experiments have confirmed the frequency of interaction between pairs and sometimes groups of three genes (discussed above). However, the incidence of pair-wise interactions implies that at some sites more than three related genes might interact within common transcription factories (Schoenfelder et al. 2010). To date, technical limitations have hampered our ability to define the extent of inter-chromosomal interactions within gene networks. However dynamically promiscuous gene domains might turn out to be, our present understanding of chromatin dynamics, the structure of chromosome territories and the organization of the synthetic centres within transcription factories suggest that the upper limit for local gene interaction will be around ten. Whether such network interactions lead to stable repositioning of chromatin loops once the primary interaction are formed or can dynamically refold towards the dissociated ground state once transcription terminates remains to be confirmed – to date the dynamic relocation of specific gene loci has not been visualized in living cells. Despite such questions, it is clear that the analysis of transcription-dependent gene associations will yield surprises for many years to come.

7.6 Global Nuclear Organization

Many of the features discussed above imply that nuclei are ordered and that sites such as factories (Cook 2010), speckles (Lamond and Spector 2003) and even transcription factor hotspots are organized in nuclear space by interaction with structural nuclear elements that in extracted cells are revealed as the nuclear matrix (Berezney et al. 1996) or nucleoskeleton (Cook 1999). Such a structured view of nuclear organization does, however, appear to contradict the possibility that large chromatin domains, and in rare cases entire chromosomes, might be able to engage new active sites during changes/reprogramming of gene expression. The movement of a complete chromosome during interphase would represent an unprecedented level of nuclear remodelling that normally would only be seen during cell division. How to resolve the apparent conflict between chromatin dynamics and stable nuclear structures is a difficult question, which justifies our efforts to understanding the molecular principles that drive nuclear structure and function.

A useful way of approaching this question is to think about nuclei as containing two major compartments: the chromatin-rich and inter-chromatin domains (Lanctôt et al. 2007). The chromatin-rich compartment typically occupies 40–50% of the nucleus in a proliferating mammalian cell. DNA within this compartment is folded into a series of higher-order structures, with DNA foci that contain roughly 1 Mbp of DNA representing a major feature of both structure and function (Cremer and Cremer 2001). The local organization of foci within chromosome territories defines a preferred steady state for the local organization, though it is important to realise that this can be quite plastic, so that the structure of a region may vary from cell to cell (Shopland et al. 2006). Mammalian genomes have been known for many years to be organized at even higher levels into to chromosome bands that have variable sizes, mostly in the range 2–20 Mbp of DNA. These bands are defined in broad terms by their functional properties, with chromosomal R-bands being gene-rich and transcriptionally active and G-bands relatively gene-poor and more often synthetically inert. Notably, the distribution of foci within the active and inactive compartments is characteristic and distinct (Goetze et al. 2007; Fig. 7.2). Key points to be noted in these typical foci is that they are folded locally to occupy sub-domains within the relevant chromosome territory – the fact that territories approximate to ellipsoids in structure presumably reflects an innate features of the way their constituent parts interact. Most notably, chromatin domains within active chromosomal regions are more dispersed that those in inert chromatin, so that active regions typically occupy ~3–5-fold large volumes. In addition, while the foci with euchromatin domains are known to be locally dynamic, the mobility of foci in heterochromatin is restricted by local clustering of foci (Fig. 7.2). This organization also drives a local polarization of CTs, so that in broad terms the active chromatin domains located towards the nuclear interior and inactive ones towards the nuclear periphery.

There is compelling evidence that active sites of nuclear function are associated with the nucleoskeleton (Cook 1999) and related nuclear matrix (Berezney et al. 1996). While the nuclear and cytoplasmic compartments contain proteins with common structural properties their components and functions are clearly unique. The cytoskeleton is composed of actin nicrofilaments, myosin microtubles and a network of intermediate filaments. Though these components have specific roles the networks interact to form an integrated functional entity. Members of the relevant protein families are also found in nuclei. Nuclear specific variants of actin and myosin have been described (de Lanerolle et al. 2005). However, while these proteins have similar functions to their cytoskeletal counterparts, with fundamental roles in nuclear dynamics (Dundr et al. 2007; Hu et al. 2008), there is to date no compelling evidence that these proteins are present within extensive filament networks. The nuclear intermediate filaments (IF) are a nuclear specific form of class V IF that is populated by the nuclear lamin proteins (Gruenbaum et al. 2003). Four major lamin proteins (and minor

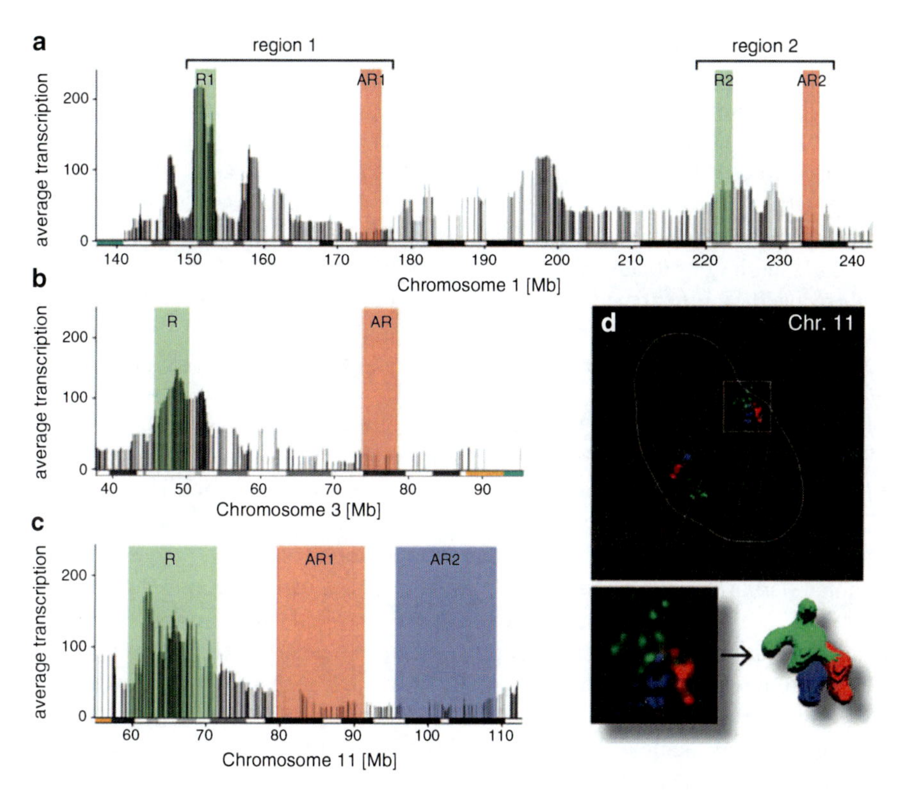

Fig. 7.2 DNA foci – units of higher-order chromatin folding. DNA in mammalian nuclei folds into chromatin domains that contain ~1 Mbp of DNA. Across the human genome, genes are clustered into gene-rich domains – called gene ridges (R) or gene islands – and separated by gene-poor domains – called anti-ridges (AR) or gene deserts. These correspond with regions of euchromatin that form chromosomal R-bands and heterochromatin that form G-bands, respectively. Three loci from chromosomes 1, 3 and 11, were selected by bio-informatic analysis based on the local distribution of the adjacent gene-rich and gene-poor domains (**a**–**c**). The nuclear distribution of 3 regions of chromosome 11 (one ridge and two anti-ridges) highlighted in (**c**) were visualized using fluorescent in situ hybridization to probe the locus (**d**) – probes covering the three regions were differentially labelled in order to reveal their distribution within the typical nucleus shown (**d**). Based on the lengths of the labelled regions (Mbp in **a**–**c**) and the number of labelled foci (coloured spots in (**d**), see high magnification view below), each DNA focus contains ~1Mbp of DNA. Note that the gene-rich and gene-poor domains are discrete and that the gene-rich domains occupy a larger volume that gene-poor domains that have a very similar DNA content (Reproduced from Goetze et al. (2007) and published with permission of the American Society of Microbiology)

splice variants) are found in human cells: lamins A, B1, B2 and C. As in the cytoskeletal IF networks, the nuclear lamin proteins are known to form extended filaments with similar structural properties to the cytoskeletal filaments. The lamin network is most elaborate along the inner face of the nuclear membrane and there is some evidence that an internal but diffuse network provides the core structure of the nucleoskeleton.

Lamin proteins have been shown to support a number of essential nuclear roles (Dechat et al. 2008). The most widely agreed role is in preserving the integrity of the nuclear compartment, by maintaining the shape and mechanical stability of the nucleus. Most notably, cells deficient in expression of A-type lamins, as well as mutants expressed in some laminopathies (Dauer and Worman 2009) – notably Emery-Dreifuss muscular dystrophy – display severely impaired mechanotransduction and reduced mechanical stiffness. There is evidence also that the structure of the nucleoskeleton contributes to global integrity of the structural cellular networks, with reduced lamin expression leading to defects in cell migration (Dechat et al. 2008). Phenotypes seen in other laminopathies imply that normal expression of the nuclear lamin proteins is required to maintain global nuclear architecture and normal patterns of gene expression for some genes. Rare premature ageing syndromes (Ding and Shen 2008) such as Hutchinson-Gilford progeria syndrome (HGPS) provides a most intriguing example of how defect in behaviour of lamin proteins can have profound consequences for nuclear organization and function. In the most prevalent mutation seen in this syndrome, a single nucleotide substitution (1824 C>T) in *LMNA* results in activation of a cryptic splice site, which generates a mutant lamin A protein that lack 50 amino acids at its C terminus. This mutant protein has compromised processing with abnormal membrane association, disorganization of the nuclear periphery, particularly regarding the structure of the heterochromatin compartment, and alterations in nuclear shape. In addition to these structural changes, progeria cells express profound mitotic defects, with delayed cytokinesis and nuclear reassembly, defective chromosome segregation and formation of binucleate cells (Cao et al. 2007).

As well as these structural and related organizational nuclear roles the nuclear lamin has also been implicated in essential nuclear functions (see Dechat et al. 2008 for review). Relevant to gene expression, initial studies showed that a dominant negative mutant of lamin A was able to disrupt the structure of the lamin network and leave cells with severely compromised RNA synthesis. Later work used RNA interference to deplete the different lamin proteins and showed that loss of lamin B1, but not B2 or the A-type lamins, resulted is severe transcriptional defects (Tang et al. 2008). Notably, the deterioration in assembled lamin B1 correlated first with the decline is transcription by RNA polymerase II and later loss of RNA polymerase I activity within nucleoli (Tang et al. 2008; Martin et al. 2009). In the latter case, falling lamin B1 concentrations correlated with the gradual deconstruction of the active transcription centres within nucleoli and isolation of purified nucleoli showed these active centres to contain lamin B1 under normal conditions. The interpretation of these studies was that nuclear structures that contain lamin proteins and essentially lamin B1 are required for the assembly of active sites of RNA synthesis, perhaps because they provide structures that contribute to the organization of the active sites.

Many proteins that interact both directly and indirectly with the nuclear lamin networks define the functional properties of the lamin proteins (Schirmer and Foisner 2007). Many such proteins have been characterised, and key players such as emerin, lamin B receptor, BAF and the lamin associated proteins LAP1 and

LAP2 α/β are known to play fundamental roles in global nuclear and chromatin organization. Different lamin-associated proteins regulate lamin interactions at the nuclear periphery and within the nuclear interior. Analysis of chromatin domains that are able to interact with the nuclear periphery provides an interesting insight into major interactions at the nuclear edge. The approach used is to tag an ectopically expressed lamin or lamin-associated protein with a bacterial DNA methylase (Guelen et al. 2008). The methylase is expressed in the vicinity of the assembled lamin filaments so that any chromatin that comes into close contact will be methylated. As the methylation is specific to the bacterial enzyme used there is no relevant demethylase, so that the novel Me residues provide a stable memory of DNA regions that have interacted with the lamina; to build a comprehensive picture of the entire interactome samples are generally analysed after expressing the methyalse for many hours. In human cells, hotspot of lamin interaction correlate broadly with heterochromatin regions, showing that in the preferred steady-state organization heterochromatin is located in the vicinity of the most lamin-rich nuclear domains, at the nuclear periphery. Interestingly, the lamin-associated chromatin domains also reveal clues about the structural constraints that define the different chromatin compartments. In particular, the boundaries of the lamin associated domains can be shown to represent transition regions between the lamin associated domains that contain a very low density of the boundary forming chromatin factor CTCF (Guelen et al. 2008). In recent years, the insulator protein CTCF has emerged as a good candidate to define boundary elements that punctuate the genome to form higher-order chromatin domains (West and Fraser 2005; Phillips and Corces 2009). Intriguingly, sites of CTCF binding have also been shown to be sites of cohesin accumulation, suggesting that they might assume special structural properties that contribute to architecture of chromatin loops (Hadjur et al. 2009). In addition, hotspots of CTCF binding have been shown to establish unique features in the local chromatin environment, which might contribute to the formation of higher-order chromatin conformations.

7.7 Modelling Gene Expression

As our knowledge of nuclear organization continue to evolve it becomes imperative to explore how our understanding of the links between nuclear structure and function can contribute to any desire to develop holistic understanding of gene expression programmes, particularly in light of developing ideas about gene expression networks.

From the complexities discussed above, it is clear that the development of anything approaching a holistic understanding of nuclear function will be an immense task that will inevitably involve the development of novel in silico tools and sophisticated modelling approaches. For higher eukaryotes, present models of gene expression are at a preliminary developmental stage. Good examples of the early models centre of key regulators of cell proliferation/death, notably p53

(Geva-Zatorsky et al. 2006), nuclear factor-kappaB (NF-κB; Nelson et al. 2004; Ashall et al. 2009) and the cell cycle machinery (Conradie et al. 2010). The NF-κB network is an especially exciting system as the signalling and linked gene expression network interact with other networks that regulate proliferation, cell cycle, genome stability and apoptosis. Though the NF-κB network is one of the most complex mammalian systems for which a predictive model has been developed the model is still limited by information on the relative contribution of different features of nuclear organization. The present models are based on available parameters of the network components such as concentration, kinetic constants and association/dissociation rates and focus on the behaviour of the signalling network based on the ability of NF-κB complexes to interact with target genes following induction. In cells prior to induction, NF-κB is retained in the cytoplasm through its interaction with the inhibitor of κB (IκB). Following signal induction, a membrane-associated receptor complex is activated and this results in activation of Iκkinase (IκK), which then phosphorylates targets within the IκB-NF-κB complex. IκB is targeted for degradation and the NF-κB then moves to the nucleus to activate target genes. For genes that are tagged with suitable reporter, the activation of gene expression can be monitored by protein expression and mathematical back-calculations used to simulate events at individual promoters. A key feature of the published models (Nelson et al. 2004; Ashall et al. 2009) is the events that cause the signalling to switch off. In this system, a number of feedback loops result from the activity of NF-κB target genes. Three proteins, IκBα, IκBε and A20 are activated at different times by NF-κB. The IκB isoforms turn off signalling by recruiting NF-kB back into the cytoplasm, switching of NF-κB signalling once sufficient copies of the inhibitor proteins have been made (human cells have about 65,000 copies of NF-κB so expression of the required number of inhibitor molecules is estimated to require at least 1 h). The A20 induction loop feeds back into the signal transduction pathway as an inhibitor of IκK. Together these three feedback loops provide a integrated off switch which because of its complexity has the added value of generating biological heterogeneity into the network, which could be an important factor in modulating cell physiology.

Present models incorporate signalling kinetic and concentrations of the key signalling components but have many missing features that will modulate behaviour of the system. For example, there is a profound lack of understanding as to how different post-translational modifications to NF-κB components – there are five transcription factors that can form different homo- and heterodimers and these are differentially modified and stabilized into response to different signalling inputs – influence the spectrum of target gene expression. There is only limited information on how the chromatin environment and changes that occur at different target gene promoters might influence the process of activating gene expression and for most target genes (there are at least 300) the pathway and timing of recruitment of factors to the promoters is unknown. In addition, we have no reliable information on how higher-order chromatin architecture influences the patterns of gene activation and if the levels of expression involve the formation of interacting gene expression networks, as described above. Finally, direct visualization and quantification of nascent transcripts in single cells is needed to reveal the precision with which transcription is activated at specific target genes and define the need to incorporate deterministic or stochastic features into the modelling framework. From this simplistic appraisal of the modelling activities it is clear that our ability to think about nuclei as complex but highly integrated machines is still extremely limited. Even so, the recognition of these limitations, based on recent advances in understanding nuclear organization, will inevitable stimulate the discovery of ever more exciting developments for many years to come.

References

Ashall L, Horton CA, Nelson DE et al (2009) Pulsatile stimulation determines timing and specificity of NF-kappaB-dependent transcription. Science 324:242–246

Baxter J, Merkenschlager M, Fisher AG (2002) Nuclear organization and gene expression. Curr Opin Cell Biol 14:372–376

Berezney R, Mortillaro MJ, Ma H et al (1996) The nuclear matrix: a structural milieu for genomic function. Int Rev Cytol 162A:1–65

Berger SL (2007) The complex language of chromatin regulation during *transcription*. Nature 447:407–412

Bernstein BE, Meissner A, Lander ES (2007) The mammalian epigenome. Cell 128:669–681

Brown JM, Green J, Das Neves RP et al (2008) Association between active genes occurs at nuclear speckles and is modulated by chromatin environment. J Cell Biol 182:1083–1097

Cao K, Capell BC, Erdos MR et al (2007) A lamin A protein isoform overexpressed in Hutchinson-Gilford progeria syndrome interferes with mitosis in progeria and normal cells. Proc Natl Acad Sci U S A 104:4949–4954

Conradie R, Bruggeman FJ, Ciliberto A et al (2010) Restriction point control of the mammalian cell cycle via the cyclin E/Cdk2:p27 complex. FEBS J 277:357–367

Cook PR (1999) The organization of replication and transcription. Science 284:1790–1795

Cook PR (2010) A model for all genomes: the role of transcription factories. J Mol Biol 395:1–10

Cremer T, Cremer C (2001) Chromosome territories, nuclear architecture and gene regulation in mammalian cells. Nat Rev Genet 2:292–301

Dauer WT, Worman HJ (2009) The nuclear envelope as a signaling node in development and disease. Dev Cell 17:626–638

Dechat T, Pfleghaar K, Sengupta K et al (2008) Nuclear lamins: major factors in the structural organization and function of the nucleus and chromatin. Genes Dev 22:832–853

de Lanerolle P, Johnson T, Hofmann WA (2005) Actin and myosin I in the nucleus: what next? Nat Struct Mol Biol 12:742–746

Ding SL, Shen CY (2008) Model of human aging: recent findings on Werner's and Hutchinson-Gilford progeria syndromes. Clin Interv Aging 3:431–444

Dundr M, Ospina JK, Sung MH et al (2007) Actin-dependent intranuclear repositioning of an active gene locus in vivo. J Cell Biol 179:095–103

Emerson BM (2002) Specificity of gene regulation. Cell 109:267–270

Eskiw CH, Rapp A, Carter DR, Cook PR (2008) RNA polymerase II activity is located on the surface of protein-rich transcription factories. J Cell Sci 121:1999–2007

Ferrai C, Xie SQ, Luraghi P et al (2010) Poised transcription factories prime silent uPA gene prior to activation. PLoS Biol 8:e1000270

Filipski E, Lévi F (2009) Circadian disruption in experimental cancer processes. Integr Cancer Ther 8:298–302
Fraser P, Bickmore W (2007) Nuclear organization of the genome and the potential for gene regulation. Nature 447:413–417
Geva-Zatorsky N, Rosenfeld N, Itzkovitz S et al (2006) Oscillations and variability in the p53 system. Mol Syst Biol 2:0033
Goetze S, Mateos-Langerak J, Gierman HJ et al (2007) The three-dimensional structure of human interphase chromosomes in related to the transcriptome map. Mol Cell Biol 27:4475–4487
Gruenbaum Y, Goldman RD, Meyuhas R et al (2003) The nuclear lamina and its functions in the nucleus. Int Rev Cytol 226:1–62
Guelen L, Pagie L, Brasset E et al (2008) Domain organization of human chromosomes revealed by mapping of nuclear lamina interactions. Nature 453:948–951
Hadjur S, Williams L, Ryan M et al (2009) Cohesins form chromosomal cis-interactions at the developmentally regulated IFNG locus. Nature 460:410–413
Hager GL, McNally JG, Misteli T (2009) Transcription dynamics. Mol Cell 35:741–753
Hu Q, Kwon YS, Nunez E et al (2008) Enhancing nuclear receptor-induced transcription requires nuclear motor and LSD1-dependent gene networking in interchromatin granules. Proc Natl Acad Sci U S A 105:19199–19204
Jackson DA, Hassan AB, Errington RJ et al (1993) Visualization of focal sites of transcription within human nuclei. EMBO J 12:1059–1065
Jackson DA, Iborra FJ, Manders EM et al (1998) Numbers and organization of RNA polymerases, nascent transcripts, and transcription units in HeLa nuclei. Mol Biol Cell 9:1523–1536
Jackson DA, Pombo A (1998) Replicon clusters are stable units of chromosome structure: evidence that nuclear organization contributes to the efficient activation and propagation of S phase in human cells. J Cell Biol 140:1285–1295
Ko CH, Takahashi JS (2006) Molecular components of the mammalian circadian clock. Hum Mol Genet 15:R271–277
Lamond AI, Spector DL (2003) Nuclear speckles: a model for nuclear organelles. Nat Rev Mol Cell Biol 4:605–612
Lanctôt C, Cheutin T, Cremer M et al (2007) Dynamic genome architecture in the nuclear space: regulation of gene expression in three dimensions. Nat Rev Genet 8:104–115
Lieberman-Aiden E, van Berkum NL, Williams L et al (2009) Comprehensive mapping of long-range interactions reveals folding principles of the human genome. Science 326:289–293
Maniatis T, Reed R (2002) An extensive network of coupling among gene expression machines. Nature 416:499–506
Martin C, Chen S, Maya-Mendoza A et al (2009) Lamin B1 maintains the functional plasticity of nucleoli. J Cell Sci 122:1551–1562
Matsuo T, Yamaguchi S, Mitsui S et al (2003) Control mechanism of the circadian clock for timing of cell division in vivo. Science 302:255–259
Misteli T (2007) Beyond the sequence: cellular organization of genome function. Cell 128:787–800
Nelson DE, Ihekwaba AE, Elliott M et al (2004) Oscillations in NF-kappaB signaling control the dynamics of gene expression. Science 306:704–708
Noordermeer D, de Laat W (2008) Joining the loops: beta-globin gene regulation. IUBMB Life 60:824–833
Osborne CS, Chakalova L, Brown KE et al (2004) Active genes dynamically colocalize to shared sites of ongoing transcription. Nat Genet 36:1065–1071
Osborne CS, Chakalova L, Mitchell JA et al (2007) Myc dynamically and preferentially relocates to a transcription factory occupied by Igh. PLoS Biol 8:e192
Papantonis A, Wada Y, Ohta Y et al. (2010) Changing contacts between TNFα-responsive genes point to immobilization of active RNA polymerases (Submitted)
Paschos GK, Baggs JE, Hogenesch JB et al (2010) The role of clock genes in pharmacology. Annu Rev Pharmacol Toxicol 50:187–214

Patrinos GP, de Krom M, de Boer E et al (2004) Multiple interactions between regulatory regions are required to stabilize an active chromatin hub. Genes Dev 18:1495–1509
Phillips JE, Corces VG (2009) CTCF: Master weaver of the genome. Cell 137:1194–1211
Schirmer EC, Foisner R (2007) Proteins that associate with lamins: many faces, many functions. Exp Cell Res 313:2167–2179
Schoenfelder S, Sexton T, Chakalova L et al (2010) Preferential associations between co-regulated genes reveal a transcriptional interactome in erythroid cells. Nat Genet 42:53–61
Shopland LS, Lynch CR, Peterson KA et al (2006) Folding and organization of a contiguous chromosome region according to the gene distribution pattern in primary genomic sequence. J Cell Biol 174:27–38
Tang CW, Maya-Mendoza A, Martin C et al (2008) The integrity of a lamin B1-dependent nucleoskeleton is a fundamental determinant of RNA synthesis in human cells. J Cell Sci 121:1014–1024
Volpi EV, Chevret E, Jones T et al (2000) Large-scale chromatin organization of the major histocompatibility complex and other regions of human chromosome 6 and its response to interferon in interphase nuclei. J Cell Sci 113:1565–1576
Wada Y, Ohta Y, Xu M et al (2009) A wave of nascent transcription on activated human genes. Proc Natl Acad Sci U S A 106:18357–18361
West AG, Fraser P (2005) Remote control of gene transcription. Hum Mol Genet 14:R101–R111
Wood PA, Yang X, Hrushesky WJ (2009) Clock genes and cancer. Integr Cancer Ther 8:303–308

Index